THE ORGANIC CHEM LAB SURVIVAL MANUAL

A STUDENT'S GUIDE TO TECHNIQUES

THIRD EDITION

JAMES W. ZUBRICK

Hudson Valley Community College

John Wiley & Sons, Inc.
New York   Chichester   Brisbane   Toronto   Singapore

Acquisitions Editor	Nedah Rose
Marketing Manager	Catherine Faduska
Production Supervisor	Sandra Russell
Digital Production Supervisor	Joanne Kelman
Manufacturing Manager	Lorraine Fumuso/Andrea Price
Copyediting Supervisor	Elizabeth Swain

This book was set in New Century Schoolbook by
Digital Production and printed and bound by Courier Stoughton.
The cover was printed by Phoenix.

Library of Congress Cataloging-in-Publication Data

Zubrick, James W.
    The organic chem lab survival manual: a student's guide to
techniques / James W. Zubrick.--3rd ed.
    Includes index.
    ISBN 0-471-57504-6
    1. Chemistry, Organic--Laboratory manuals.   I. Title.
QD261.Z83   1992   92-24585
547'.0078--0078   CIP

Printed in the United States of America

10   9   8   7   6   5   4   3

To Cindy

PREFACE TO THE
THIRD EDITION

It is heartening to hear of your book being read and enjoyed, literally cover to cover, by individuals ranging from talented high-school science students to Professors Emeritus of the English language. Even better to hear that you have a chance to improve that book, based upon the above comments, comments by reviewers, and the experience gained from working with the text.

The goal of this edition is, as it has been from the first edition, to present the basic techniques in the organic chemistry laboratory with the emphasis on doing the work correctly the first time. As always, I have relied on the unfailingly thoughtful and helpful comments of users to guide me in deciding what changes needed to be made for this third edition.

The most notable additions are sections detailing the use of microscale equipment, something that is becoming the scale of choice, and includes: the conical vial and its handling; the O-ring-seal cap; gas collection apparatus; Pasteur pipet tips such as filtering, extraction, and washing; handling syringes and needles; the rubber septum; Craig tube recrystallization; analytical balance technique for the organic laboratory; and Hickman still distillations. In addition, readers will find short new sections on the operation of the FTIR, and waste disposal. Numerous other sections have been revised based on the editorial comments of reviewers.

Many people deserve credit for their assistance in producing this edition: I would also like to thank

William Dailey
    University of Pennsylvania
Rudolph Goetz
    Michigan State University
Marco Pagnotta
    Barnard College
Kathleen Petersen
    University of Notre Dame
Frederick Heldrich
    College of Charleston

for their valuable comments and suggestions in making this edition more useful for students of organic chemistry laboratory.

Finally, I'd like to thank Ms. Nedah Rose, Chemistry Editor at John Wiley & Sons, Inc., especially for her help in the determination of what should be what with this edition, and Ms. Karen Kosztolnyik, editorial assistant, for help in keeping information, that tended to get away from me, straight. Ms. Sandra Russell, Senior Production Supervisor, Ms. Elizabeth Swain, Senior Copyeditor, and Ms. Madelyn Lesure, Design Director deserve a great deal of credit for helping to put this third edition together in such a professional manner.

J.W. Zubrick
Hudson Valley Community College
June 10, 1992

SOME NOTES ON STYLE

It is common to find instructors railing against poor usage and complaining that their students cannot do as much as to write one clear, uncomplicated, communicative English sentence. Rightly so. Yet I am astonished that the same people feel comfortable with the long and awkward passive voice, the pompous "we" and the clumsy "one," and that damnable "the student," to whom exercises are left as proofs. These constructions, which appear in virtually all scientific texts, do *not* produce clear, uncomplicated, communicative English sentences. And students do learn to write, in part, by following example.

I do not go out of my way to boldly split infinitives, nor do I actively seek prepositions to end sentences with. Yet by these constructions alone, I may be viewed by some as aiding the decline in students' ability to communicate.

E.B. White, in the second edition of *The Elements of Style* (Macmillan, New York, 1972, p. 70), writes

> Years ago, students were warned not to end a sentence with a preposition; time, of course, has softened that rigid decree. Not only is the preposition acceptable at the end, sometimes it is more effective in that

spot than anywhere else. "A claw hammer, not an axe, was the tool he murdered her with." This is preferable to "A claw hammer, not an ax, was the tool with which he murdered her."

Some infinitives seem to improve on being split, just as a stick of round stovewood does. "I cannot bring myself to really like the fellow." The sentence is relaxed, the meaning is clear, the violation is harmless and scarcely perceptible. Put the other way, the sentence becomes stiff, needlessly formal. A matter of ear.

We should all write as poorly as White.

With the aid of William Strunk and E.B. White in *The Elements of Style* and that of William Zinsser in *On Writing Well*, Rudolph Flesch in *The ABC of Style,* I have tried to follow some principles of technical communication lately ignored in scientific texts: use the first person, put yourself in the reader's place, and, the best for last, use the active voice and a personal subject.

The following product names belong to the respective manufacturers. Registered trademarks are indicated here, as appropriate; in the text, the symbol is omitted.

Corning®	Corning Glass Works, Corning, New York
Drierite®	W.A. Hammond Drierite Company, Xenia, Ohio
Fisher-Johns®	Fisher Scientific Company, Pittsburgh, Pennsylvania
Luer-Lok®	Becton, Dickinson and Company, Rutherford, New Jersey
Mel-Temp®	Laboratory Devices, Cambridge, Massachusetts
Millipore®	Millipore Corporation, Bedford, Massachusetts
Swagelok®	Crawford Fitting Company, Solon, Ohio
Teflon®	E.I DuPont de Nemours & Company, Wilmington, Delaware
Variac®	General Radio Company, Concord, Massachusetts

CONTENTS

SAFETY FIRST, LAST, AND ALWAYS

The organic chemistry laboratory is potentially one of the most dangerous of undergraduate laboratories. That is why you must have a set of safety guidelines. It is a very good idea to pay close attention to these rules, for one very good reason:

The penalties are only too real.

Disobeying safety rules is not at all like flouting many other rules. *You can get seriously hurt*. No appeal. No bargaining for another 12 points so you can get into medical school. Perhaps as a patient, but certainly not as a student. So, go ahead. Ignore these guidelines. But remember—

You have been warned!

1. ***Wear your goggles.*** Eye injuries are extremely serious and can be mitigated or eliminated if you keep your goggles on *at all times*. And I mean *over your eyes*, not on top of your head or around your neck. There are several types of eye protection available, some of it acceptable, some not, according to local, state, and federal laws. I like the clear plastic goggles that leave an unbroken red line on your face when you remove them. Sure, they fog up a bit, but the protection is superb. Also, think about getting chemicals or chemical fumes trapped under your contact lenses before you wear them to lab. Then don't wear them to lab. Ever.

2. ***Touch not thyself.*** Not a Biblical injunction, but a bit of advice. You may have just gotten chemicals on your hands, in a concentration that is not noticeable, and sure enough, up go the goggles for an eye wipe with the fingers. Enough said.

3. ***There is no "away."*** Getting rid of chemicals is a very big problem. You throw them from here, and they wind up poisoning someone else. Now there are some laws to stop that from happening. The rules were really designed for industrial waste, where there are hundreds of gallons of waste that have the same composition. In a semester of organic lab there will be much smaller amounts of different materials. Waste containers could be provided for everything, but this is not practical. If you don't see the waste can you need, ask your instructor. When in doubt, *ask*.

4. ***Bring a friend.*** *You must never work alone*. If you have a serious accident and you are all by yourself, you might not be able to get help before you die. Don't work alone, and don't work at unauthorized times.

5. *Don't fool around.* Chemistry is serious business. Don't be careless or clown around in lab. You can hurt yourself or other people. You don't have to be somber about it; just serious.

6. *Drive defensively.* Work in the lab as if someone else were going to have an accident that might affect you. Keep the goggles on because *someone else* is going to point a loaded, boiling test tube at you. *Someone else* is going to spill hot, concentrated acid on your body. Get the idea?

7. *Eating, drinking, or smoking in lab.* Are you kidding? Eat in a chem lab? Drink in a chem lab??? Smoke, and blow yourself up????

8. *Keep it clean.* Work neatly. You don't have to make a fetish out of it, but try to be *neat*. Clean up spills. Turn off burners or water or electrical equipment when you're through with them.

9. *Where it's at.* Learn the location and proper use of the fire extinguishers, fire blankets, safety showers, and eyewash stations.

10. *Making the best-dressed list.* Keep yourself covered from the neck to the toes—no matter what the weather. That might include long-sleeved tops that also cover the midsection. Is that too uncomfortable for you? How about a chemical burn to accompany your belly-button, or an oddly shaped scar on your arm in lieu of a tatoo. And pants that come down to the shoes, covering any exposed ankles is probably a good idea as well. No open-toed shoes, sandals, or canvas-covered footwear. No loose-fitting cuffs on the pants or the shirts. Nor are dresses appropriate for lab, guys. Keep the midsection covered. Tie back that long hair. And a small investment in a lab coat can pay off, projecting that extra professional touch. It gives a lot of protection too. Consider wearing disposable gloves. Clear polyethylene ones are inexpensive, but the smooth plastic is slippery, and there's a tendency for the seams to rip open when you least expect it. Latex examination gloves keep the grip and don't have seams, but they cost more. Gloves are not perfect protectors. Reagents like bromine can get through and cause severe burns. They'll buy you some time though and can help mitigate or prevent severe burns. Oh, yes—laboratory aprons: they only cover the *front,* so your exposed legs are still at risk from behind.

11. *Hot under the collar.* Many times you'll be asked or told to heat something. Don't just automatically go for the Bunsen burner. That way lies *fire.* Usually—

No Flames!

Try a hot plate, try a heating mantle (see Chapter 18, "Sources of Heat"), but try to stay away from flames. Most of the fires I've had to put out started when some bozo decided to heat some flammable solvent in an open beaker. Sure, there are times when you'll **HAVE** to use a flame, but use it away from all flammables and in a hood (Fig. 1), and only with the permission of your instructor.

12. ***Work in the hood.*** A hood is a specially constructed workplace that has, at the least, a powered vent to suck noxious fumes outside. There's also a safety glass or plastic panel you slide down as protection from exploding apparatus (Fig. 1). If it is at all possible, treat every chemical (even solids) as if toxic or bad-smelling fumes came from it, and carry out as many of the operations in the organic lab as you can *inside a hood*, unless told otherwise.

13. ***Keep your fingers to yourself.*** Ever practiced "finger chemistry"? You're unprepared so you have a lab book out, and your finger points to the start of a sentence. You move your finger to the end of the first line and do that operation—

"Add this solution to the beaker containing the ice-water mixture"

And WHOOSH! Clouds of smoke. What happened? The next line reads—

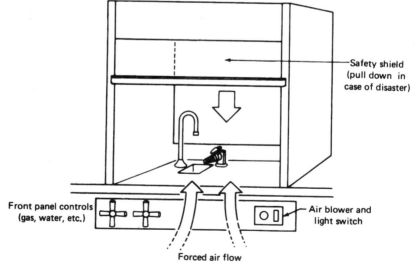

Fig. 1   A typical hood.

"very carefully as the reaction is highly exothermic"

But you didn't read that line, or the next, or the next. So you are a danger to yourself and everyone else. Read and take notes on any experiment before you come to lab (see Chapter 2, "Keeping a Notebook").

14. ***What you don't know can hurt you.*** If you are not sure about an operation, or you have any question about handling anything, *please* ask your instructor before you go on. Get rid of the notion that asking questions will make you look foolish. Following this safety rule may be the most difficult of all. Grow up. Be responsible for yourself and your own education.

15. ***Blue Cross or Blue Shield?*** Find out how you would get medical help, if you needed it. Sometimes during a summer session, the school infirmary is closed and you would have to be transported to the nearest hospital.

These are a few of the safety guidelines for an organic chemistry laboratory. You may have others particular to your own situation.

ACCIDENTS WILL NOT HAPPEN

That's an attitude you might hold while working in the laboratory. You are **not** going to do anything, or get anything done to you, that will require medical attention. If you do get cut, and the cut is not serious, wash the area with water. If there's serious bleeding, apply direct pressure with a clean, preferably sterile, dressing. For a minor burn, let cold water run over the burned area. For chemical burns to the eyes or skin, flush the area with lots of water. In every case, get to a physician if at all possible.

If you have an accident, *tell your instructor immediately. Get help!* This is no time to worry about your grade in lab. If you put grades ahead of your personal safety, be sure to see a psychiatrist after the internist finishes.

DISPOSING OF WASTE

Once you do your reaction, since your mother probably doesn't take organic lab with you, you'll have to clean up after yourself. I've hesitated to write this section for a very long time because the rules for cleaning up vary greatly according to, but not limited to, Federal, state, and local laws, as well as individual practices at individual colleges. There are even differences—legally—if you or your instructor do the cleaning up. And, as always, things do seem to run to money—the more money you have to spend, the more you can throw away. So there's not much point to even trying to be authoritative about waste disposal in this little manual, but there are a few things I have picked up that you should pay attention to. Remember, my classification scheme may not be the same as the one you'll be using. When in doubt, **ask! Don't just throw everything into the sink. Think.**

Note to the picky: The word nonhazardous, as applied here, means relatively benign, as far as organic laboratory chemicals go. After all, even pure water, carelessly handled, can kill you.

1. *Nonhazardous insoluble waste.* Paper, corks, sand, alumina, silica gel, sodium sulfate, magnesium sulfate, and so on can probably go into the ordinary wastebaskets in the lab. Unfortunately, these things can be contaminated with hazardous waste (see discussion below), and then they need special handling.

2. *Nonhazardous soluble solid waste.* Some organics, such as benzoic acid, are relatively benign and can be dissolved with a lot of tap water and flushed down the drains. But if the solid is that benign, it might just as well go out with the nonhazardous insoluble solid waste, no? Check with your instructor; watch out for contamination with more hazardous materials.

3. *Nonhazardous soluble liquid waste.* Plain water, down the drains, as well as water-soluble substances not otherwise covered below. Ethanol can probably be sent down the drains, but butanol? It's not that water soluble, so it probably should go into the general organic waste container. Check with your instructor; watch out for contamination with more hazardous materials.

4. *Nonhazardous insoluble liquid waste.* Compound such as 1-butanol (discussed above), diethyl ether, and most other solvents and compounds not covered otherwise. In short, this is the traditional "organic waste" category.

5. *Generic hazardous waste.* Pretty much all else not listed separately. Hydrocarbon solvents (hexane, toluene), amines (aniline, triethylamine), amides, esters, acid chlorides, and on and on. Again,

traditional "organic waste." Watch out for incompatibilities, though, before you throw just anything in any waste bucket. If the first substance in the waste bucket was acetyl chloride and the second was diethylamine (both hazardous liquid wastes), the reaction may be quite spectacular. You may have to use separate hazardous waste containers for these special circumstances.

6. ***Halogenated organic compounds.*** 1-Bromobutane and t-butyl chloride, undergraduate laboratory favorites, should go into their own waste container as "halogenated hydrocarbons." There's a lot of agreement on this procedure for these simple compounds. But what about your organic unknown, 4-bromobenzoic acid? I'd have you put it, and any other organic with a halogen, in the "halogenated hydrocarbon" container and not flush it down the drain as a harmless organic acid, as you might do with benzoic acid.

7. ***Strong inorganic acids and bases.*** Neutralize them, dilute them, and flush them down the sink. At least as of this writing.

8. ***Oxidizing and reducing agents.*** Reduce the oxidants and oxidize the reductants before disposal. Be careful! Such reactions can be highly exothermic. Check with your instructor before proceeding.

9. ***Toxic heavy metals.*** Convert to a more benign form, minimize the bulk, and put in a separate container. If you do a chromic acid oxidation, you might reduce the more hazardous Cr^{+6} to Cr^{+3} in solution and then precipitate the Cr^{+3} as the hydroxide, making lots of expensive-to-dispose-of chromium solution into a tiny amount of solid precipitate. There are some gray areas, though. Solid manganese dioxide waste from a permanganate oxidation should probably be considered a hazardous waste. It can be converted to a soluble Mn^{+2} form, but should Mn^{+2} go down the sewer system? I don't know the effect of Mn^{+2} (if any) in the environment. But do we want it out there?

Mixed Waste

Mixed waste has its own special problems and raises even more questions. Here are some examples:

1. ***Preparation of acetominophen (Tylenol): A multi-step synthesis.*** You've just recrystallized 4-nitroaniline on the way to acetominophen, and washed and collected the product on your Buchner funnel. So you

have about 30–40 mL of this really orange solution of 4-nitroaniline and byproducts. The nitroaniline is very highly colored, the byproducts probably more so, so there isn't really a lot of solid organic waste in this solution; not more than perhaps a hundred milligrams or so. Does this go down the sink, or is it treated as organic waste? Remember, you have to package, label, and transport to a secure disposal facility what amounts to 99.9% perfectly safe water. Check with your instructor.

2. ***Preparation of 1-bromobutane.*** You've just finished the experiment and you're going to clean out your distillation apparatus. There is a residue of 1-bromobutane coating the three-way adapter, thermometer, inside of the condenser, and the adapter at the end. Do you wash the equipment in the sink and let this minuscule amount of a halogenated hydrocarbon go down the drain? Or do you rinse everything with a little acetone into yet another beaker and pour that residue into the "halogenated hydrocarbon" bucket, fully aware that most of the liquid is acetone and doesn't need special halide treatment? Check with your instructor.

3. ***The isolation and purification of caffeine.*** You've dried a methylene chloride extract of caffeine and are left with methylene chloride saturated drying agent. Normally a nonhazardous solid waste, no? Yes. But where do you put it while the methylene chloride is on it? Some would have you put it in a bucket in a hood and let the methylene chloride evaporate into the atmosphere. Then the drying agent is nonhazardous solid waste. But you've merely transferred the problem somewhere else. Why not just put the whole mess in with the "halogenated hydrocarbons"? Usually "halogenated hydrocarbons" go to a special incinerator equipped with traps to remove HCl or HBr produced upon burning. Drying agents don't burn very well, and the cost of shipping the drying agent part of this waste is very high.

Question: Is this cost higher than that of letting the methylene chloride evaporate into the atmosphere? The current and past Republican administrations see fit to keep studying the interactions of pollutants in the atmosphere rather than actually do anything that might disturb "business as usual," so this method of pollution control is, at least, highly patriotic. What should you do? Again, ask your instructor.

In these cases, as in many other questionable situations, I tend to err on the side of caution and consider the bulk of the waste to have the attributes of its most hazardous component. This is, unfortunately, the most expensive way to look at the matter. In the absence of guidelines:

1. Don't make a lot of waste in the first place.

2. Make it as benign as possible. (Remember though, such reactions can be highly exothermic, so proceed with caution.)

3. Reduce the volume as much as possible.

Oh. Try to remember that sink drains can be tied together, and if you pour a sodium sulfide solution down one sink, while someone else is diluting an acid in another sink, toxic, gagging, rotten-egg-smelling hydrogen sulfide can back up the drains in your entire lab, and maybe even the building.

KEEPING A
NOTEBOOK

A **research notebook** is perhaps one of the most valuable pieces of equipment you can own. With it you can duplicate your work, find out what happened at leisure, and even figure out where you blew it. General guidelines for a notebook are:

1. The notebook must be bound permanently. No loose leaf or even spiral-bound notebooks will do. It should have a sewn binding so that the only way pages can come out is to cut them out. (8 ½ x 11 in. is preferred). **Duplicate carbon notebooks** are available that let you make a carbon copy that is removable and that you can hand in. And if the pages aren't already numbered, you might want to do it yourself.

2. *Use waterproof ink! Never pencil!* Pencil will disappear with time, and so will your grade. Cheap ink will wash away and carry your grades down the drain. Never erase! Just draw *one* line through ~~yuor errers~~ your errors so that they can still be seen. And never, never, never cut any pages out of the notebook!

3. Leave a few pages at the front for a table of contents.

4. Your notebook is your friend, your confidant. Tell it:

    a. What you have done. Not what it says to do in the lab book. What you, yourself, have done.

    b. Any and *all* observations: color changes, temperature rises, explosions…, anything that occurs. Any *reasonable* explanation *why* whatever happened, happened.

5. Skipping pages is *extremely* poor taste. It is **NOT** done!

6. List the IMPORTANT chemicals you'll use during each reaction. You should include USEFUL **physical properties:** the name of the compound, molecular formula, molecular weight, melting point, boiling point, density, and so on. You might have entries for the number of moles and notes on handling precautions. Useful information, remember. The *CRC Handbook of Chemistry and Physics*, originally published by the Chemical Rubber Company and better known as the *CRC Handbook*, is one place to get this stuff (see Chapter 3, "Interpreting a Handbook").

Note the qualifier "USEFUL," if you can't use any of the information given, do without it! You look things up *before* the lab so you can tell what's staring back out of the flask at you during the course of the reaction.

Your laboratory experiments can be classified as either of two major types: a technique experiment or a synthesis experiment. Each requires different handling.

A TECHNIQUE EXPERIMENT

In a technique experiment, you get to practice a certain operation *before* you have to do it in the course of a synthesis. Distilling a mixture of two liquids to separate them is a typical technique experiment.

Read the following handwritten notebook pages with some care and attention to the *typeset* notes in the margin. A thousand words are worth a picture or so (Figs. 2-4).

Notebook Notes

1. Use a descriptive title for your experiment. *Distillation.* This implies you've done *all* there is in the *entire* field of distillation. You haven't? Perhaps all you've done is *The Separation of a Liquid Mixture by Distillation.* Hmmmmmm.
2. Writing that first sentence can be difficult. Try stating the obvious.
3. There are no large blank areas in your notebook. Draw sloping lines through them. Going back to enter observations after the experiment is over is *not professional.* Initial and date pages anytime you write anything in your notebook.
4. Note the appropriate changes in verb tense. Before you do the work, you might use the present or future tense when you write about something that *hasn't happened yet.* During the lab, since you are supposed to write what you've actually done just after the time you've actually done it, a simple past tense is sufficient.

A SYNTHESIS EXPERIMENT

In a synthesis experiment, the point of the exercise is to prepare a clean sample of the product you want. All the operations in the lab (e.g., distillation, recrystallization) are just means to this end. The preparation of 1-bromobutane is a classic synthesis and is the basis of the next series of handwritten notebook pages.

Explanatory
title

Numbered
page

The Separation of a Liquid Mixture by Distillation

9/13/86

6

This is the
Saturday before
lab.

Distillation is one of the methods of separation and purification of liquids. We will be given an unknown liquid mixture and will have to separate it by distillation.

It's often hard
to start. Hint:
state the obvious.

After we get the unknown, should *immediately* ~~dry it with~~ dry the liquid over anhydrous magnesium sulfate. The setup is as detailed in the laboratory manual with some changes:

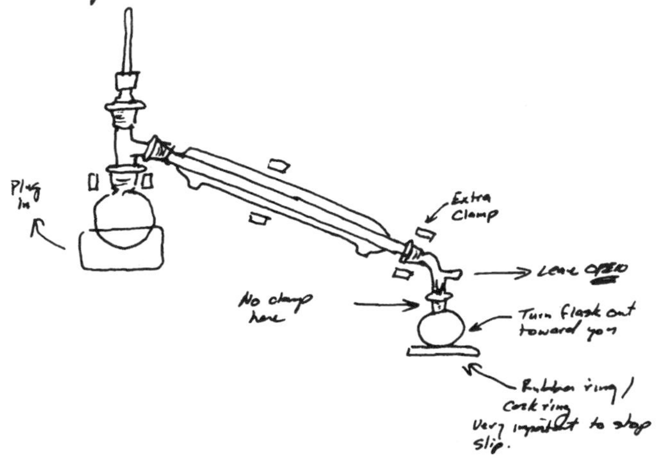

Plug In

Extra Clamp

Leave OPEN

No clamp here

Turn flask out toward you

Rubber ring / Cork rim Very important to stop slip.

Local procedure
change, probably
from handout.

We will be using Thermowell heaters and will not need Variacs. Vacuum adapter clamped at angle, rotated toward me in order to make it easy to change flasks

9/13/86

Fig. 2 Notebook entry for a technique experiment (1).

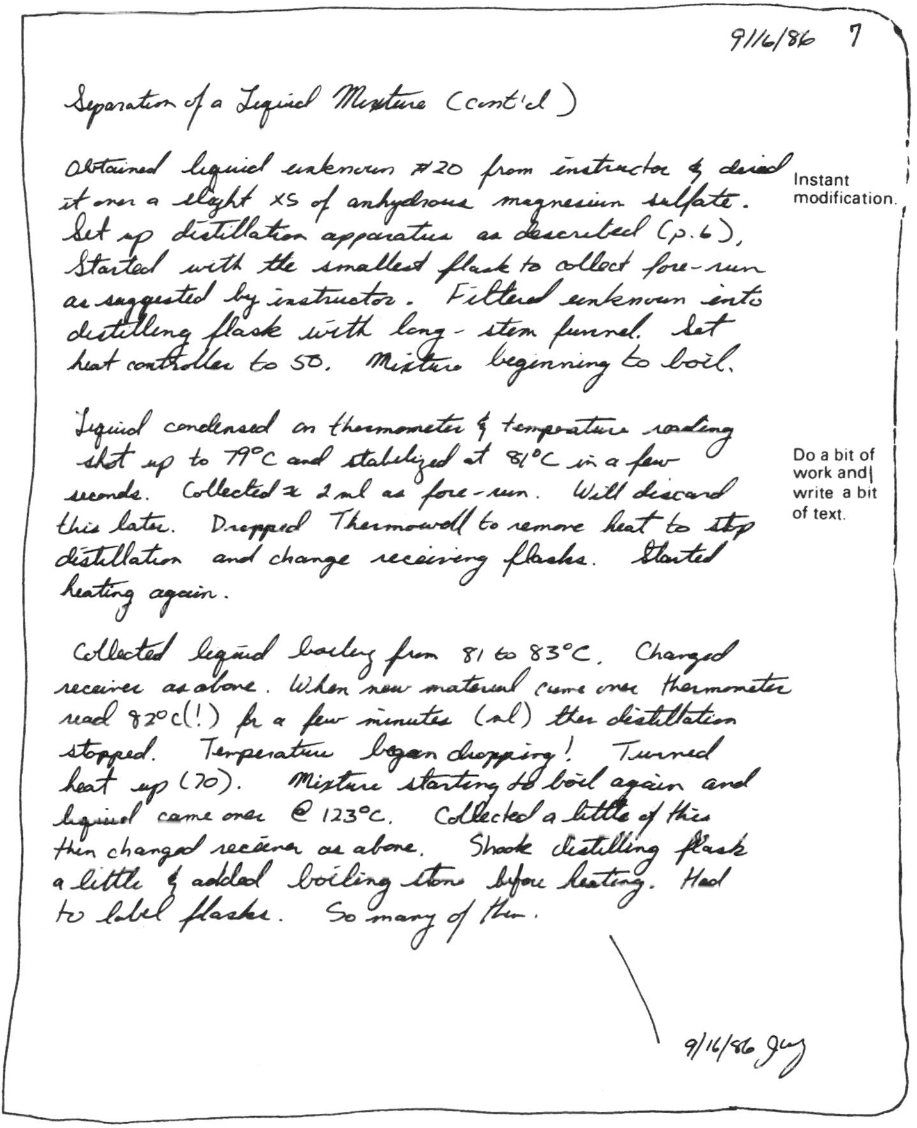

2

9/16/86   7

Separation of a Liquid Mixture (cont'd)

Obtained liquid unknown #20 from instructor & dried
it over a slight xs of anhydrous magnesium sulfate.
Set up distillation apparatus as described (p.6),
Started with the smallest flask to collect fore-run
as suggested by instructor. Filtered unknown into
distilling flask with long-stem funnel. Set
heat controller to 50. Mixture beginning to boil.

Instant modification.

Liquid condensed on thermometer & temperature reading
shot up to 79°C and stabilized at 81°C in a few
seconds. Collected ≈ 2 ml as fore-run. Will discard
this later. Dropped Thermowell to remove heat to stop
distillation and change receiving flasks. Started
heating again.

Do a bit of work and write a bit of text.

Collected liquid boiling from 81 to 83°C. Changed
receiver as above. When new material came over thermometer
read 82°C (!) for a few minutes (al) then distillation
stopped. Temperature began dropping! Turned
heat up (70). Mixture starting to boil again and
liquid came over @ 123°C. Collected a little of this
then changed receiver as above. Shook distilling flask
a little & added boiling stone before heating. Had
to label flasks. So many of them.

9/16/86 Guy

Fig. 3   Notebook entry for a technique experiment (2).

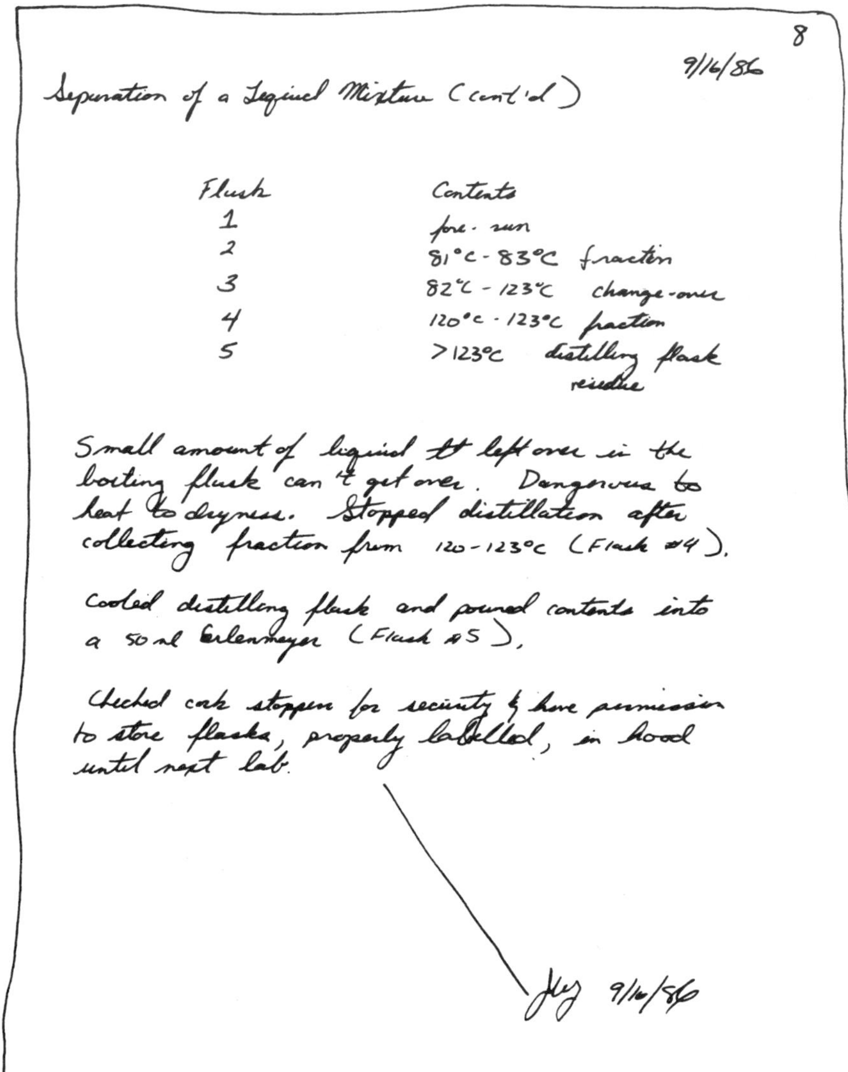

8

9/16/86

Separation of a Liquid Mixture (cont'd)

Flask	Contents
1	fore-run
2	81°C - 83°C fraction
3	82°C - 123°C change-over
4	120°C - 123°C fraction
5	>123°C distilling flask residue

Small amount of liquid lt left over in the boiling flask can't get over. Dangerous to heat to dryness. Stopped distillation after collecting fraction from 120-123°C (Flask #4).

Cooled distilling flask and poured contents into a 50 ml Erlenmeyer (Flask #5).

Checked cork stoppers for security & have permission to store flasks, properly labelled, in hood until next lab.

JLy 9/16/86

Fig. 4   Notebook entry for a technique experiment (3).

Pay careful attention to the typeset notes in the margins, as well as the handwritten material. Just for fun, go back and see how much was written for the distillation experiment, and note how that is handled in this synthesis (Figs. 5-10).

Once again, if your own instructor wants anything different, do it. The art of notebook keeping has many schools—follow the perspective of your own.

Notebook Notes

1. Use a descriptive title for your experiment. *n-Butyl Bromide*. So what? Did you drink it? Set it on fire? What?! *The Synthesis of 1-Bromobutane from 1-Butanol*—now *that's* a title.
2. Do you see a section for unimportant side reactions? No. Then don't include any.
3. In this experiment, we use a 10% aqueous sodium hydroxide solution as a wash (see Chapter 15, "Extraction and Washing") and anhydrous calcium chloride as a drying agent (see Chapter 10, "Drying Agents"). These are not listed in the Table of Physical Constants. They are neither reactants nor products. Every year, however, somebody always lists the physical properties of *solid* sodium hydroxide, calcium chloride drying agent, and a bunch of other reagents that have nothing to do with the main synthetic reaction. I'm especially puzzled by the listing of solid sodium hydroxide in place of the 10% solution.
4. **Theoretical yield** (not yeild) calculations always seem to be beyond the ken of a lot of you, even though these are exercises right out of the freshman year chemistry course. Yes, we do expect you to remember some things from courses past, the least of which is where to look this up. I've put a sample calculation in the notebook (Fig. 6), that gets the mass (g) of the desired product (1-bromobutane) from the volume (ml) of one reactant (1-butanol). Why from the 1-butanol and not from the sulfuric acid or sodium bromide? It's the 1-butanol we are trying to convert to the bromide, and we use a **molar excess** (often abbreviated XS) of everything else. The 1-butanol is, then, the **limiting reagent:** the reagent present in the smallest molar ratio. Note the use of the density to get from volume to mass (ml to g), molecular weight to go from mass to number of moles (g to mol), the stoichiometric ratio (here 1:1) to get moles of product from moles of limiting reagent, and, finally, reapplication of molecular weight to get the mass (g) of the product.

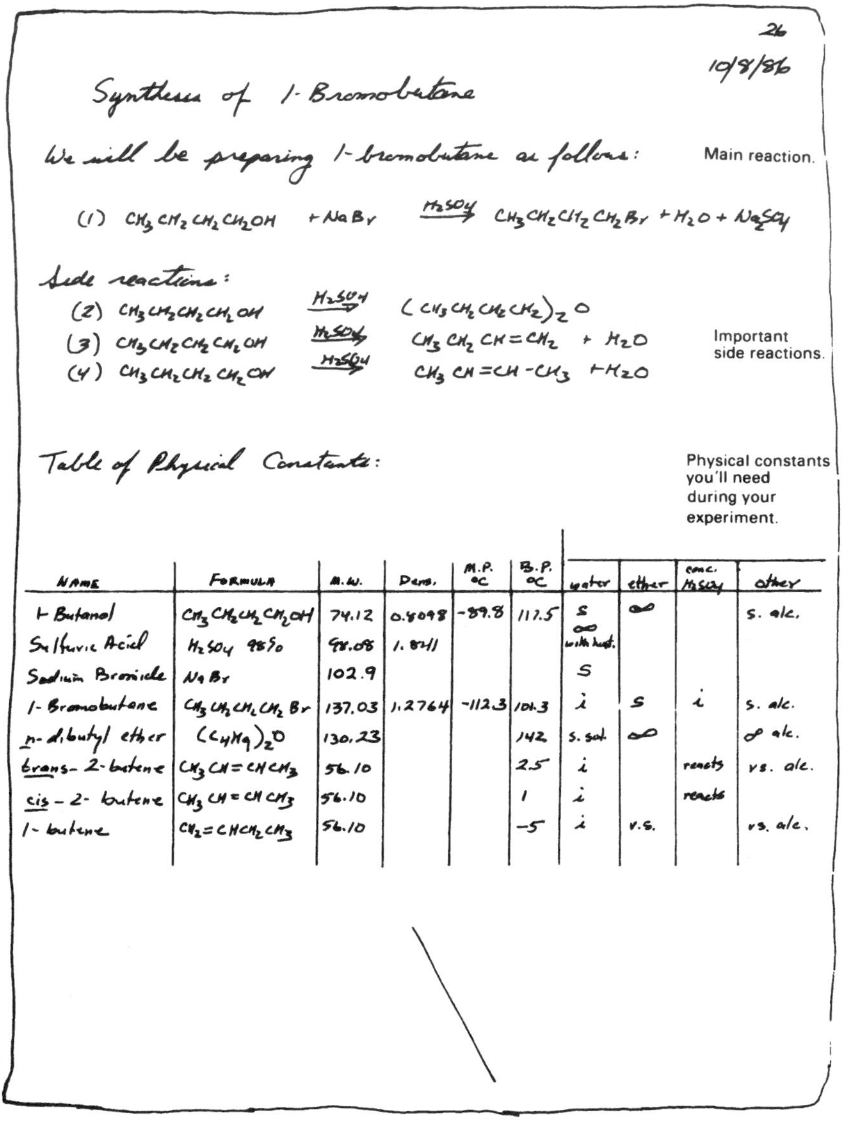

26

10/8/86

Synthesis of 1-Bromobutane

We will be preparing 1-bromobutane as follows: **Main reaction.**

(1) $CH_3 CH_2 CH_2 CH_2 OH$ + NaBr $\xrightarrow{H_2SO_4}$ $CH_3 CH_2 CH_2 CH_2 Br$ + H_2O + Na_2SO_4

Side reactions:

(2) $CH_3 CH_2 CH_2 CH_2 OH$ $\xrightarrow{H_2SO_4}$ $(CH_3 CH_2 CH_2 CH_2)_2 O$

(3) $CH_3 CH_2 CH_2 CH_2 OH$ $\xrightarrow{H_2SO_4}$ $CH_3 CH_2 CH = CH_2$ + H_2O **Important side reactions.**

(4) $CH_3 CH_2 CH_2 CH_2 OH$ $\xrightarrow{H_2SO_4}$ $CH_3 CH = CH - CH_3$ + H_2O

Table of Physical Constants:

Physical constants you'll need during your experiment.

NAME	FORMULA	M.W.	Dens.	M.P. °C	B.P. °C	water	ether	conc. H₂SO₄	other
1-Butanol	$CH_3 CH_2 CH_2 CH_2 OH$	74.12	0.8098	-89.8	117.5	S ∞	∞		s. alc.
Sulfuric Acid	H_2SO_4 98%	98.08	1.841			with heat.			
Sodium Bromide	NaBr	102.9				S			
1-Bromobutane	$CH_3 CH_2 CH_2 CH_2 Br$	137.03	1.2764	-112.3	101.3	i	s	i	s. alc.
n-dibutyl ether	$(C_4H_9)_2 O$	130.23			142	s. sol	∞		∞ alc.
trans-2-butene	$CH_3 CH = CH CH_3$	56.10			2.5	i		reacts	vs. alc.
cis-2-butene	$CH_3 CH = CH CH_3$	56.10			1	i		reacts	
1-butene	$CH_2 = CH CH_2 CH_3$	56.10			-5	i	v.s.		vs. alc.

Fig. 5   Notebook entry for a synthesis experiment (1).

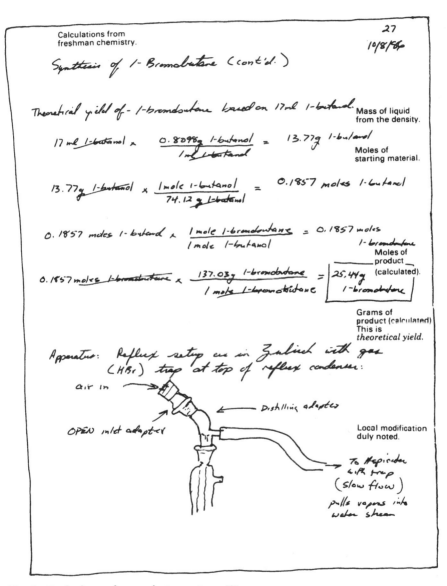

Fig. 6 Notebook entry for a synthesis experiment (2).

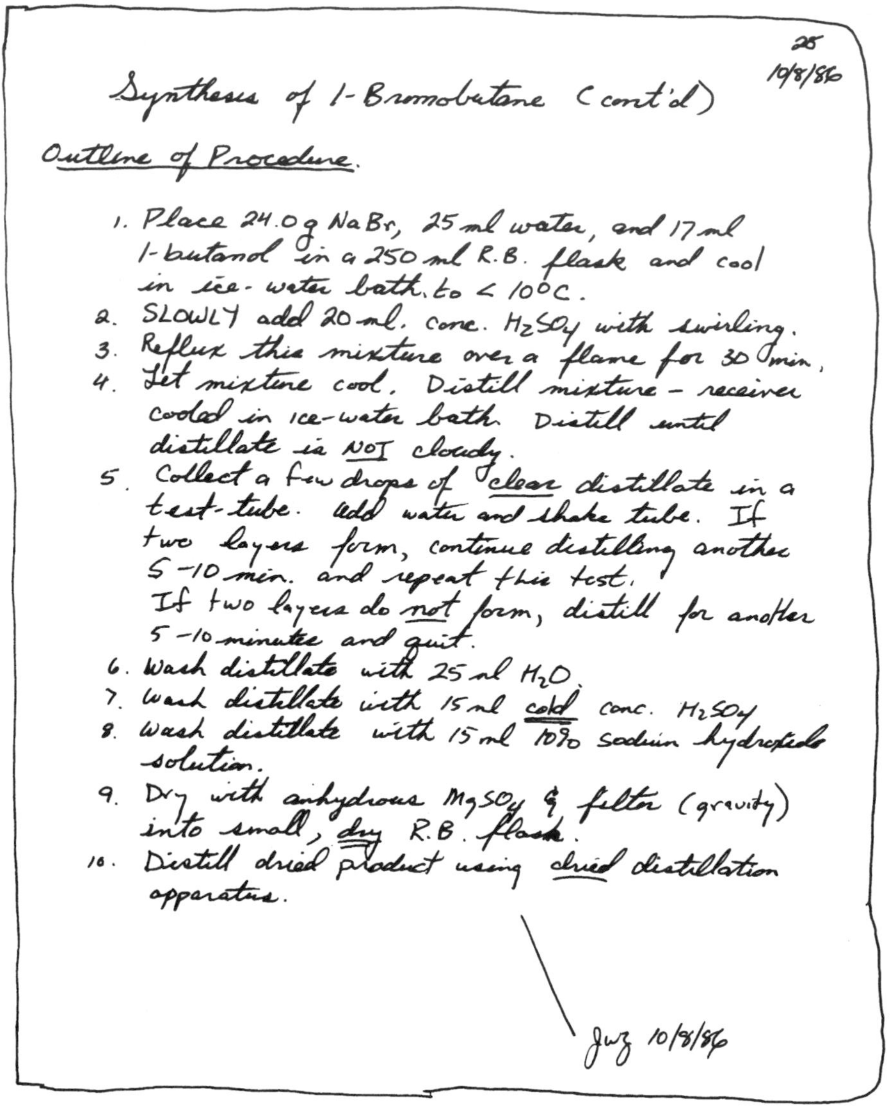

25
10/8/86

Synthesis of 1-Bromobutane (cont'd)

Outline of Procedure.

1. Place 24.0 g NaBr, 25 ml water, and 17 ml 1-butanol in a 250 ml R.B. flask and cool in ice-water bath. to < 10°C.
2. SLOWLY add 20 ml. conc. H_2SO_4 with swirling.
3. Reflux this mixture over a flame for 30 min.
4. Let mixture cool. Distill mixture - receiver cooled in ice-water bath. Distill until distillate is NOT cloudy.
5. Collect a few drops of clear distillate in a test-tube. Add water and shake tube. If two layers form, continue distilling another 5-10 min. and repeat this test.
   If two layers do not form, distill for another 5-10 minutes and quit.
6. Wash distillate with 25 ml H_2O.
7. Wash distillate with 15 ml cold conc. H_2SO_4.
8. Wash distillate with 15 ml 10% sodium hydroxide solution.
9. Dry with anhydrous $MgSO_4$ & filter (gravity) into small, dry R.B. flask.
10. Distill dried product using dried distillation apparatus.

Jwz 10/8/86

Fig. 7 Notebook entry for a synthesis experiment (3).

2

29
10/9/86

Synthesis of 1-Bromobutane (cont'd.)

Placed 24.0 g NaBr, 25 ml water and 17 ml 1-butanol in a 250 ml R.B. flask & let cool in an ice-bath. When liquid reached 5°C, added H_2SO_4 with swirling. The mixture warmed up and turned a yellow color.

Set up for reflux with gas trap as on p.27. Mixture darkened as reflux continued.

Next step performed while reflux continues.

Placed 15 ml H_2SO_4 in erlenmeyer clamped to cool in ice-water bath for later.

After refluxing 30 min, let reaction mixture cool to room temp, then put in ice-water bath. There are two distinct layers in the flask, both an orange color. Color may be due to free bromine. <u>One</u> of the two layers is product.

Only a phrase recalls the distillation.

Distilled the mixture, and collected everything that came over up to 100°C. Initially, cloudy, white liquid came over (water + organic product?) then clear liquid. Stopped heating dutifuly flask; removed receiver and replaced it with test-tube in beaker with ice & water. Heated to distill over a few drops of liquid. Added a bit more than an equal amount of water — shook tube. <u>NO LAYERS FORMED</u>! Replaced receiving flask & distilled for 5 min. more.

Poured distillate into a 125 ml erlenmeyer separatory funnel & added 25 ml water. Water went into upper layer — upper layer is aqueous; <u>lower</u> is organic product. <u>Save lower layer</u>.

Note the recorded observations.

Fig. 8 Notebook entry for a synthesis experiment (4).

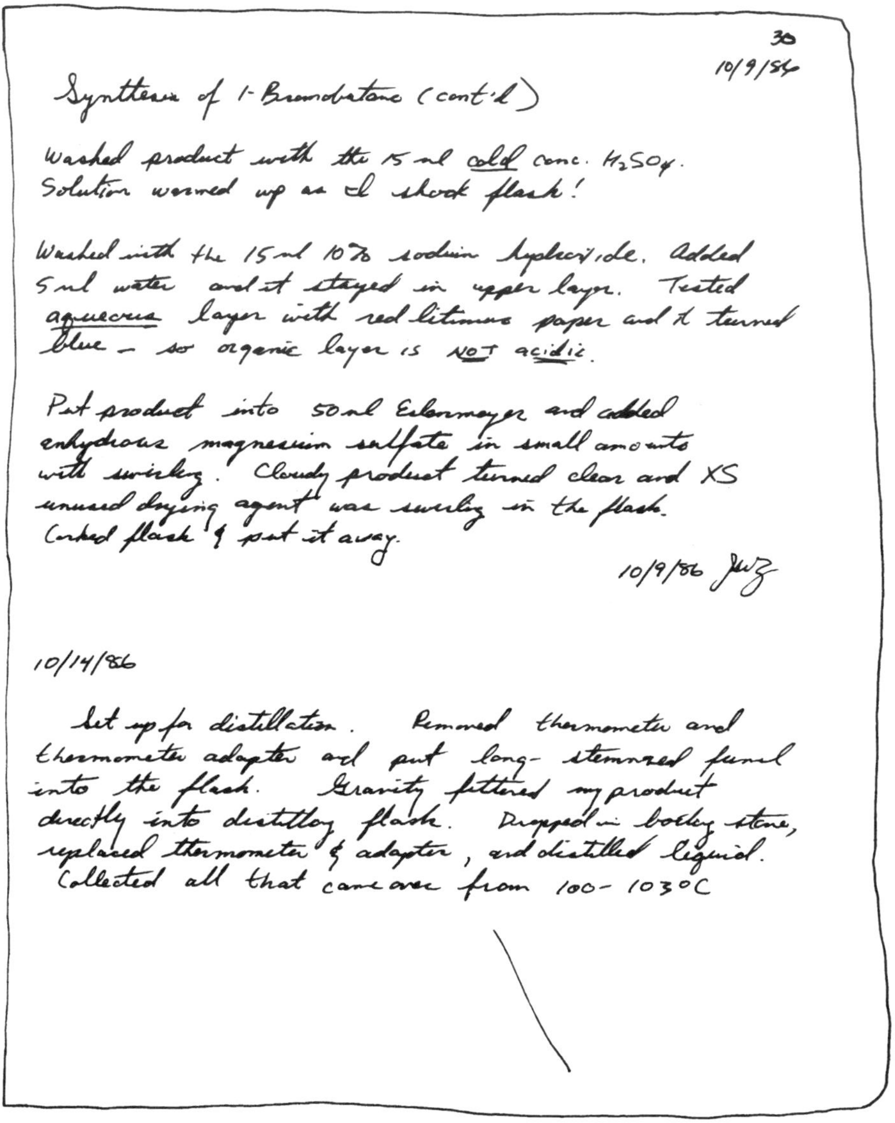

30
10/9/86

Synthesis of 1-Bromobutane (cont'd)

Washed product with the 15 ml cold conc. H₂SO₄.
Solution warmed up as I shook flask!

Washed with the 15 ml 10% sodium hydroxide. Added
5 ml water and it stayed in upper layer. Tested
aqueous layer with red litmus paper and it turned
blue — so organic layer is NOT acidic.

Put product into 50 ml Erlenmeyer and added
anhydrous magnesium sulfate in small amounts
with swirling. Cloudy product turned clear and XS
unused drying agent was swirling in the flask.
Corked flask & put it away.

10/9/86 JWZ

10/14/86

Set up for distillation. Removed thermometer and
thermometer adapter and put long-stemmed funnel
into the flask. Gravity filtered my product
directly into distilling flask. Dropped in boiling stone,
replaced thermometer & adapter, and distilled liquid.
Collected all that came over from 100-103°C

Fig. 9 Notebook entry for a synthesis experiment (5).

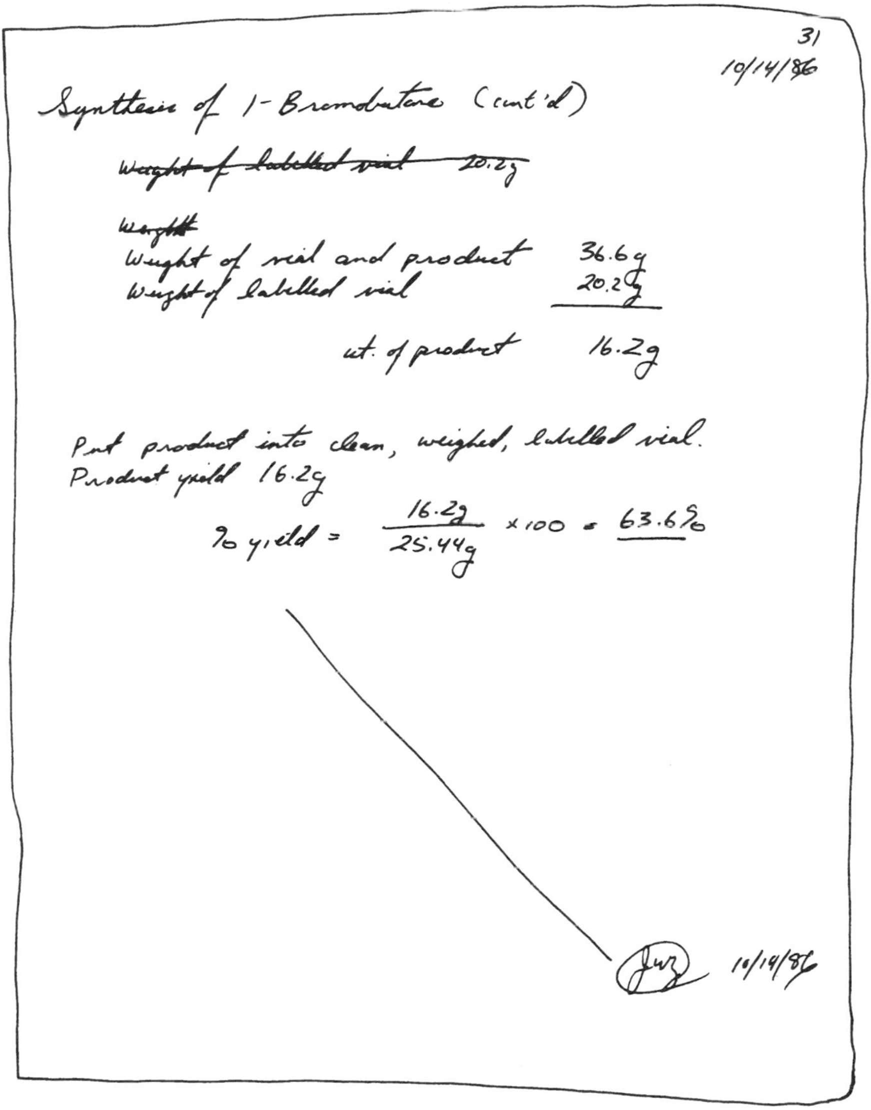

31
10/14/86

Synthesis of 1-Bromobutane (cont'd)

~~Weight of labelled vial 20.2g~~

~~Weight~~

Weight of vial and product	36.6g
Weight of labelled vial	20.2g
wt. of product	16.2g

Put product into clean, weighed, labelled vial.
Product yield 16.2g

$$\% \text{ yield} = \frac{16.2g}{25.44g} \times 100 = \underline{63.6\%}$$

JWZ 10/14/86

Fig. 10 Notebook entry for a synthesis experiment (6).

Note that this mass is *calculated*. It is NOT anything we've actually produced. In theory, we get this much. That is theoretical yield.

5. I'm a firm believer in the use of units, factor-label method, dimensional analysis, whatever you call it. I know I've screwed up if my units are (g 1-butanol)^2/mole 1-butanol.

6. Remember the huge writeup on the *Separation of a Liquid Mixture by Distillation*, drawings of apparatus and all? Well, the line "the mixture was purified by distillation" (Fig. 9) is all you write for the distillation during this synthesis.

7. At the end of the synthesis, you calculate the **percent yield**. Just divide the amount you *actually prepared* by the amount you calculated you'd get, and multiply this fraction by 100. For this synthesis, I *calculated* a yield of 25.44 g of product. For this reaction on the bench, I *actually* obtained 16.2 g of product. So:

$$(16.2 \text{ g}/ 25.44 \text{ g})(100) = \textbf{63.6\% yield}$$

Note that's not 63.624715321%. **Significant figures**, please. The product is weighed to one part in ten ([+| 0.1) and calculated to one part in one hundred ([+| 0.01). If the product weight can vary by [+| 0.1 g, what's the use of all those figures?

THE ACID TEST

After all this, you're still not sure what to write in your notebook? Try these simple tests at home.

1. Before lab. "Can I carry out this experiment without any lab manual?"

2. After lab. (I mean *immediately* after; none of this "I'll write my observations down later" garbage.) Ask yourself: "If this were someone else's notebook, could I duplicate the results exactly?"

If you can truthfully answer yes to these two questions, you're doing very well indeed.

INTERPRETING
A HANDBOOK

You should look up information concerning any organic chemical you'll be working with so that you know what to expect in terms of molecular weight, density, solubility, crystalline form, melting or boiling point, color, and so on. This information is kept in handbooks that should be available in the lab, if not in the library. Reading some of these handbooks is not easy, but once someone tells you what some of the fancy symbols mean, there shouldn't be a problem. Many of the symbols are common to all handbooks and are discussed only once, so read the entire section even if your handbook is different. There are at least four fairly popular handbooks, and I've included sample entries of 1-bromobutane and benzoic acid, a liquid and a solid you might come across in lab, to help explain things.

CRC HANDBOOK

(CRC Handbook of Chemistry and Physics, *CRC Press, Inc., Boca Raton, Florida.)* Commonly called "the *CRC*" as in "Look it up in the *CRC*." A very popular book; a classic. Sometimes you can get the last year's edition cheaply from the publisher, but it's usually for an order of 10 or more.

Entry: 1-Bromobutane (Fig. 11)

1. *No. 3683.* An internal reference number. Other tables in the handbook will use this number rather than the name.
2. *Name,...Butane, 1-bromo.* You get a systematic name and a formula.
3. *Mol. wt. 137.03.* The molecular weight of 1-bromobutane.
4. *Color,... ......* Dots! This implies 1-bromobutane is a colorless liquid: nothing special really.
5. *b.p. 101.6, 18.8^{30}.* The normal boiling point, at 760 torr, is 101.6°C. The 18.8 has a tiny superscript to tell you that 18.8°C is the boiling point at 30 torr.
6. *m.p. -112.4.* The melting point of solid 1-bromobutane. Handbooks report only the TOP of the melting point range. You, however, should report the *entire* range.
7. *Density. 1.2758$^{20/4}$.* Actually, this particular number is a **specific gravity**. This is a mass of the density of the liquid taken at 20°C *referred to* (divided by) the density of the same mass of water at 4°C.

PHYSICAL CONSTANTS OF ORGANIC COMPOUNDS (Continued)

No.	Name, Synonyms, and Formula	Mol. wt.	Color, crystalline form, specific rotation and λ_{max} (log ϵ)	b.p. °C	m.p. °C	Density	nD	Solubility	Ref.
2531	Benzoic acid . . . . $C_6H_5CO_2H$	122.13	mcl lf or nd	249, 133[10]	122.13	1.0749[130]	1.504[12]	al, eth, ace, bz, chl	B9[3], 360
2532	Benzoic acid, 2-acetamido . . . . 2-$(CH_3CONH)C_6H_4CO_2H$	179.18	nd (aa)	. . . .	185	1.2659[15/4]	. . . . . .	eth, ace, bz	B14[3], 922
3683	Butane, 1-bromo . . . . $CH_3CH_2CH_2CH_2Br$	137.03	. . . .	101.6, 18.8[30]	−112.4	1.2758[20/4]	1.4401[20]	al, eth, ace, chl	B1[4], 258
3684	Butane, 1-bromo-4-chloro . . . . $BrCH_2CH_2CH_2CH_2Cl$	171.48	. . . .	174–5[756], 63–4[10]	. . . . . .	1.488[20/4]	1.4885[25]	al, eth, chl	B1[4], 264

Fig. 11 Sample *CRC* entries from the 61st edition.

That's what the tiny 20/4 means. Notice the units will cancel. A number without the modifying fraction is a true density (in g/ml) at the temperature given.

8. n_D **1.4401²⁰.** This is the index of refraction (see Chapter 28, "Refractometry") obtained using the yellow light from a sodium lamp (the D line). Yes, the tiny 20 means it was taken at 20°C.

9. **Solubility. al, eth, ace, chl.** This is what 1-bromobutane must be soluble in. There are a lot of solvents, and here are the abbreviations for some of them:

al	alcohol	eth	ether
bz	benzene	chl	chloroform
peth	petroleum ether	w	water
aa	acetic acid	MeOH	methanol
lig	ligroin	CCl_4	carbon tetrachloride
to	toluene	ace	acetone

Some solvents have such a long tradition of use, they are our old friends and we use very informal names for them:

alcohol. Ethyl alcohol; ethanol.
ether. Diethyl ether; ethoxyethane.
pet. ether. Petroleum ether. Not a true ether, but a low-boiling (30–60°C) hydrocarbon fraction like gasoline.
ligroin. Another hydrocarbon mixture with a higher boiling range (60–90°C) than pet. ether.

10. **Ref. B1⁴ 258.** Reference to listing in a set of German handbooks called Beilstein. Pronounce the German "ei" like the long i, and stop yourself from saying "Beelsteen" or some such nonsense. 1-Bromobutane is in the fourth supplement (⁴), Volume 1 (B1), on page 258.

Entry: Benzoic Acid (Fig. 11)

There are a few differences in the entries, what with benzoic acid being a solid, and I'll point these out. If I don't reexamine a heading, see the explanation back in the 1-Bromobutane entry for details.

1. *Color, ...mcl lf or nd.* Monoclinic leaflets or needles. This is the shape of the crystals. There are many different crystalline shapes and colors, and I can't list them all—but here's a few:

pl	plates	mcl	monoclinic
nd	needles	rh	rhombus
lf	leaves	ye	yellow
pr	prisms	pa	pale

   I've included #2532 (benzoic acid, 2-acetamido) to show that you sometimes get a bonus. Here nd(aa) means you get needle-like crystals from acetic acid. Acetic acid (aa) is the recrystallization solvent (see Chapter 13, "Recrystallization"), and you don't have to find it on your own. Thus, pa ye nd (al) means that pale yellow needles are obtained when you recrystallize the compound from ethanol.

2. *Density. 1.0749^{130}.* This is an actual **density** of benzoic acid taken at 130°C. There is no temperature ratio as there is for the specific gravity (1.2659$^{15/4}$).

Nostalgia

I've included the entries from the 43rd and 49th editions of the *CRC* to show you that not all things improve with age.

1. *General Organization.* The 43rd and 49th editions make use of **boldface type** to list parent compounds and lighter type to list derivatives. Benzoic acid is a parent; there are many derivatives (Fig. 12). The 61st edition lists all compounds with the same weight (Fig. 11).

2. *Solubility tables.* Here the older editions *really* shine. The 43rd edition gives numerical solubility data for benzoic acid: 0.18^4, 0.27^{18}, 2.2^{75}. These are the actual solubilities, in grams of benzoic acid per 100 g of water at 4, 18, and 75°C, respectively. The butyl bromide (1-bromobutane) entry has helpful solubility indicators: **i.** *insoluble* in water; [8], *miscible* in alcohol; [8], *miscible* in diethyl ether.

Other popular abbreviations for the solubility of a compound are

s	soluble	i	insoluble
δ	slightly soluble		miscible, mixes in all proportions
h	solvent must be hot	v	very

PHYSICAL CONSTANTS OF ORGANIC COMPOUNDS (Continued)

No.	Name	Synonyms and Formula	Mol. wt.	Crystalline form, color and specific rotation	m.p. °C	b.p. °C	Density	nD	Solubility						Ref.
									w	al	eth	ace	bz	other solvents	
b1239	**Benzoic acid** . .	Benzenecarboxylic acid. 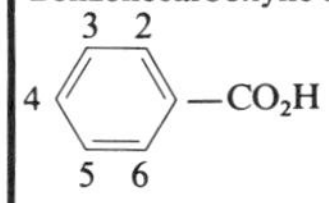 $\underset{4}{\overset{3\ \ 2}{\bigcirc}}\underset{5\ \ 6}{}-CO_2H$	122.12	mcl lf or nd	122.4	249^{760}	$1.3211^{23.3}$	1.504^{132}	δ	v	v	. .	v	CCl_4 s lig δ	**B9**^{2}, 72
b1240	—, 4-aceta-mido-pheyl ester	p-Benzoyloxyacetanilide. $C_{16}H_{13}NO_3$. *See* b1239	255.28	nd (al)	171	. . . . .	. . . . .	. . . . .	δ^h	v	. .	. .	v	aa v lig i	. . . . .
b2500	**Butane** —, **1-bromo-** * .	n-Butyl bromide. $CH_3CH_2CH_2CH_2Br$	137.03	. . . . . .	−112.3	101.3^{760}	1.2764^{20}_{4}	1.4398^{20}	i	∞	∞	. .	. .	. . . . .	**B1**^{3}, 290

Fig. 12 Sample CRC entries from the 49th edition.

What a big change from the 43rd to the 61st edition. Numerical solubility data missing, solubility indications gone, and even incomplete solubility reporting (Benzoic acid: chl, CCl$_4$, acet., me. al., bz, CS$_2$—43rd ed.; where are CCl$_4$, me. al., and CS$_2$ in the 61st?). The decrease in organizational structure, I can live with. But the new way of presenting the solubility data (what there is of it) is useless for many things you need to do in your lab. Reread the sample synthesis experiment (see Chapter 2, "Keeping a Notebook"). You need more useful solubility data for that experiment than you can extract from the most recent *CRC Handbook*. For my money, you want a fancy $17.50 doorstop, get a *CRC* 61st and up. You want useful information, get a *CRC* 60th and back. Or consult the handbook I want to talk about next.

LANGE'S

(**Lange's Handbook of Chemistry**, *McGraw-Hill Book Company, New York, New York.*) A fairly well-known but not well-used handbook. The entries are similar to those in the *CRC Handbooks,* so I'll only point out the interesting differences.

Entry: 1-Bromobutane (Fig. 14)

1. *Name. Butyl bromide(n).* Here, 1-bromobutane is listed as a substituted butyl group much like it is in the 43rd *CRC*. The systematic name is listed under synonyms.
2. *Beil. Ref. I-119.* The Beilstein reference; Volume 1, page 119, the original work (not a supplement).
3. *Crystalline form . . . lq.* It's a liquid.
4. *Specific gravity 1.275$^{20/4}$.* The tiny temperature notation is presented a bit differently, but the meaning is the same.
5. *Solubility in 100 parts water. 0.06^{16}.* 0.06 g of 1-bromobutane will dissolve in 100 g of water at 16°C. After that, no more.

Entry: Benzoic Acid (Fig. 14)

1. *Melting point. subl>= 100.* Benzoic acid starts to **sublime** (go directly from a solid to a vapor) over 100°C, before any crystals left melt at 122.4°C.

PHYSICAL CONSTANTS OF

No.	Name	Synonyms	Formula	Mol. Wt.
1449	**Benzoic acid** . . .	benzenecarboxylic acid*; phenylformic acid	C_6H_5COOH	122.12
1450	—, allyl ester. . .	allyl benzoate	$C_6H_5COOC_3H_5$ . .	162.18
1451	—, anhydride. .	See *Benzoic anhydride.*		
1452	—, benzyl ester .	benzyl benzoate; benzyl benzenecarboxylate	$C_6H_5COOCH_2C_6H_5$	212.24

*Name approved by the International Union of Chemistry.

846

PHYSICAL CONSTANTS OF

No.	Name	Synonyms	Formula	Mol. Wt.
2160	**Butyl bromide** (*n*) . .	1-bromobutane*	$CH_3(CH_2)_2CH_2Br$.	137.03
2161	*sec*-**Butyl bromide** .	2-bromobutane*; methyl-ethylbromomethane	$C_2H_5CH(CH_3)Br$.	137.03
2162	*tert*-**Butyl bromide** .	2-bromo-2-methylpropane*; trimethylbromomethane	$(CH_3)_3CBr$	137.03

*Name approved by the International Union of Chemistry.

886

Fig. 13 Sample *CRC* entries from the 41st edition.

2. *Crystalline form..., mn.pr.* Monoclinic prisms. Here, *mn* is a variant of the *mcl* abbreviation used in the *CRC*. Don't let these small differences throw you. A secret is that all handbooks have a listing of abbreviations at the front of the tables. Shhhh! Don't tell *anyone*. It's a secret.

I like the *Lange's* format, redolent of the 43rd edition of the *CRC*. The one with the useful information. The organization based on common names,

ORGANIC COMPOUNDS (Continued)

No.	Crystalline form, color and index of refraction	Density g/ml	Melting point, °C	Boiling point, °C	Solubility in grams per 100 ml of		
					Water	Alcohol	Ether, etc.
1449	col. monocl. leaf. or need., 1.53974[15]	1.2659^{15}_{4}	122	249	0.18[4] 0.27[18] 2.2[75]	47.1[15]	40[15] eth.; s. chl., CCl$_3$, acet., me. al., bz., CS$_2$
1450	yel. liq	1.058^{15}_{15}	. . . . .	230	i.	s.	∞ eth.
1451							
1452	col. oily liq., or need. or leaf., 1.5681[21]	1.114^{18}	21 (18.5)	323–4 (316–7)	i.	s.	s. eth., chl.; i. glyc.

For explanations and abbreviations see beginning of table.

847

ORGANIC COMPOUNDS (Continued)

No.	Crystalline form, color and index of refraction	Density g/ml	Melting point, °C	Boiling point, °C	Solubility in grams per 100 ml of		
					Water	Alcohol	Ether, etc.
2160	col. liq., 1.4398	1.2992^{20}_{4}	−112.4	101.6	i.	∞	∞ eth.
2161	col. liq., 1.4344[25]	1.2580^{20}_{4}	. . . . . .	91.3	i.	. . . . .	. . . .
2162	col, liq., 1.428	1.222^{30}_{4}	−20	73.3	i.	. . . . .	. . . .

For explanations and abbreviations see beginning of table.

887

rather than systematic names, can make finding an entry a bit more difficult. There's a miniature gloss at the bottom of each page to help you find related compounds.

Butyl carbinol(n), at the bottom of Fig. 14, has an index number of 404. If you're familiar with the carbinol naming scheme for alcohols, it isn't much to translate that to 1-pentanol. The entry still comes *before* the B's because Amyl alcohol(n) is another common name for 1-pentanol. On the page where 1-pentanol would show up, there's *only* a gloss entry: *1-*

Table 7-4 (*Continued*)
PHYSICAL CONSTANTS OF ORGANIC COMPOUNDS

No.	Name	Synonym	Formula	Beil. Ref.	Formula Weight
711	Benzoic acid	. . . .	$C_6H_5 \cdot CO_2H$	IX-92	122.12
712	Na salt	sodium benzoate	$C_6H_5 \cdot CO_2Na \cdot H_2O$	IX-107	162.12

Benznaphthalide 765   Benzoic acid sulfamide 5671-3

Table 7–4 (*Continued*)
PHYSICAL CONSTANTS OF ORGANIC COMPOUNDS

No.	Name	Synonym	Formula	Beil. Ref.	Formula Weight
1040	**Butyl** amine (**sec**)	. . . .	$(C_2H_5)(CH_3):CH \cdot NH_2$	IV-160	73.14
1041	amine (**iso**)	. . . .	$(CH_3)_2CH \cdot CH_2 \cdot NH_2$	IV-163	73.14
1057	bromide (**n**)	1-bromo-butane	$C_2H_5 \cdot CH_2 \cdot CH_2Br$	I-119	137.03
1058	bromide (**sec**)	2-bromo-butane	$C_2H_5 \cdot CHBr \cdot CH_3$	I-119	137.03
1059	bromide (**iso**)	1-Br-2-Me-propane	$(CH_3)_2CH \cdot CH_2Br$	I-126	137.03
1060	bromide (**tert**)	2-Br-2-Me-propane	$(CH_3)_3CBr$	I-127	137.03

Butyl borate 6117   Butyl carbinol (**iso**) 406   Butyl carbinol (**tert**) 410
Butyl carbamide 1138-9   Butyl carbinol (**sec**) 411   Butyl carbitol 2232
Butyl carbinol (**n**) 404

Fig. 14 Sample entries from *Lange's Handbook of Chemistry, 11th ed.*, J. Dean, copyright by McGraw-Hill, Inc., 1221 Avenue of the Americas, NY. Reprinted with permission.

Pentanol, 404. This brings you right back to Amyl alcohol(n). Since most textbooks and lab books are making it a *very big deal* these days to list none but the purest of pristine systematic nomenclature, you'd likely never expect the compounds to be listed this way, and that is a bit annoying. Even though you are missing out on a bit of the history in the field.

ORGANIC CHEMISTRY

Table 7-4 (*Continued*)
PHYSICAL CONSTANTS OF ORGANIC COMPOUNDS

No.	Crystalline Form and Color	Specific Gravity	Melting Point °C.	Boiling Point °C.	Solubility in 100 Parts		
					Water	Alcohol	Ether
711	mn. pr.	1.316^{25}_{4}°	122.4; subl. > 100	250.0	$0.21^{17.5°}$; $2.2^{75°}$	$46.6^{15°}$, abs. al.	$66^{15°}$
712	col. cr.	. . . . . .	$-H_2O$, 120	. . . . . .	$61^{25°}$; $77^{100°}$	$2.3^{25°}$; $8.3^{75°}$	. . . .

Benzophenone oxide 6451

ORGANIC CHEMISTRY

Table 7-4 (*Continued*)
PHYSICAL CONSTANTS OF ORGANIC COMPOUNDS

No.	Crystalline Form and Color	Specific Gravity	Melting Point °C.	Boiling Point °C.	Solubility in 100 Parts		
					Water	Alcohol	Ether
1040	col. lq.	0.724^{20}_{4}°	-104	66^{772mm}	∞	∞	∞
1041	col. lq.	0.732^{20}_{4}°	-85	68-9	∞	∞	m
1057	lq.	1.275^{20}_{4}°	-112.4	101.6	$0.06^{16°}$	∞	∞
1058	lq.	1.261^{20}_{4}°	-112.1	91.3	l.	. . . . . .	. . . .
1059	lq.	1.264^{20}_{4}°	-117.4	91.4	$0.06^{18°}$	∞	∞
1060	lq.	1.220^{20}_{4}°	-16.2	73.3, d. 210	$0.06^{18°}$	∞	m

Butyl carbonate 1892-4	Butyl citrate 6119	Butyl cyanide (*iso*) 6413
Butyl chlorocarbonate 1077-8	Butyl cyanide (*n*) 6405	Butyl cyanide (*tert*) 6246

MERCK INDEX

(**The Merck Index**, *Merck & Co., Inc., Rahway, New Jersey.*) This handbook is concerned mostly with drugs and their physiological effects. But useful information exists concerning many chemicals. Because of the

1522   *n*-Butylbenzene

1526. *n*-Butyl Bromide. *1-Bromobutane.* C_4H_9Br; mol wt 137.03. C
35.06%, H 6.62%, Br 58.32%. $CH_3(CH_2)_3Br$. Prepd from *n*-butyl alc
and a hydrobromic-sulfuric acid mixture: Kamm, Marvel, *Org. Syn.*
vol. 1, 5 (1921); Skau, McCullough, *J. Am. Chem. Soc.* 57, 2440 (1935).
 Colorless liquid. $d_4^{25}1.2686$. bp_{760} 101.3° (mp −112°). n_D^{20} 1.4398.
Insol in water; sol in alcohol, ether.

Page 216   *Consult the cross index before using this section.*

1093. Benzoic Acid. Benzenecarboxylic acid; phenylformic acid;
dracylic acid. $C_7H_6O_2$; mol wt 122.12. C 68.84%, H 4.95%, O 26.20%.
Occurs in nature in free and combined forms. Gum benzoin may
contain as much as 20%. Most berries contain appreciable amounts
(around 0.05%). Excreted mainly as hippuric acid by almost all verte-
brates, except fowl. Mfg processes include the air oxidation of toluene,
the hydrolysis of benzotrichloride, and the decarboxylation of phthalic
anhydride: Faith, Keyes & Clark's *Industrial Chemicals*, F. A. Lowen-
heim, M. K. Moran, Eds. (Wiley-Interscience, New York, 4th ed.,
1975) pp 138-144. Lab prepn from benzyl chloride: A. I. Vogel, *Practical Organic Chemistry* (Longmans, London α, 3rd ed, 1959) p 755;
from benzaldehyde: Gattermann-Wieland, *Praxis des organischen
Chemikers* (de Gruyter, Berlin, 40th ed. 1961) p 193. Prepn of ultra-
pure benzoic acid for use as titrimetric and calorimetric standard:
Schwab, Wicher, *J. Res. Nat. Bur. Standards* 25, 747 (1940). *Review:*
A. E. Williams in Kirk-Othmer *Encyclopedia of Chemical Technology*
vol. 3 (Wiley-Interscience, New York, 3rd ed., 1978) pp 778-792.

COOH

Monoclinic tablets, plates, leaflets. d 1.321 (also reported as 1.266).
mp 122.4°. Begins to sublime at around 100°. bp_{760} 249.2°; bp_{400} 277°;
bp_{200} 205.8°; bp_{100} 186.2°; bp_{60} 172.8°; bp_{40} 162.6°; bp_{20} 146.7°; bp_{10}
132.1°. Volatile with steam. Flash pt 121-131°. K at 25°: 6.40×10^{-5};
pH of satd soln at 25°: 2.8. Soly in water (g/l) at 0° = 1.7; at 10° = 2.1;

Fig. 15 Sample entries from the *Merck Index,* 10th edition.

nature of the listings, I've had to treat the explanations a bit differently
than those for the other handbooks.

Entry: 1-Bromobutane (Fig. 15)

1. ***Top of page. 1522 n-Butylbenzene.*** Just like a dictionary, each page
 has headings directing you to the first entry on that page. So, 1522 is
 not the page number but the compound number for n-Butylbenzene,

at 20° = 2.9; at 25° = 3.4; at 30° = 4.2; at 40° = 6.0; at 50° = 9.5; at 60° = 12.0; at 70° = 17.7; at 80° = 27.5; at 90° = 45.5; at 95° = 68.0. Mixtures of excess benzoic acid and water form two liquid phases beginning at 89.7°. The two liquid phases unite at the critical soln temp of 117.2°. Composition of critical mixture: 32.34% benzoic acid, 67.66% water: *see* Ward, Cooper, *J. Phys. Chem.* 34, 1484 (1930). One gram dissolves in 2.3 ml cold alc, 1.5 ml boiling alc, 4.5 ml chloroform, 3 ml ether, 3 ml acetone, 30 ml carbon tetrachloride, 10 ml benzene, 30 ml carbon disulfide, 23 ml oil of turpentine; also sol in volatile and fixed oils, slightly in petr ether. The soly in water is increased by alkaline substances, such as borax or trisodium phosphate, *see also* Sodium Benzoate.

Barium salt dihydrate, $C_{14}H_{10}BaO_4.2H_2O$, *barium benzoate.* Nacreous leaflets. *Poisonous!* Soluble in about 20 parts water; slightly sol in alc.
very sol in boiling water.

Calcium salt trihydrate, $C_{14}H_{10}CaO_4.3H_2O$, *calcium benzoate.* Orthorhombic crystals or powder. d 1.44. Soluble in 25 parts water;

Cerium salt trihydrate, $C_{21}H_{15}CeO_6.3H_2O$, *cerous benzoate.* White to reddish-white powder. Sol in hot water or hot alc.

Copper salt dihydrate, $C_{14}H_{10}CuO_4.2H_2O$, *cupric benzoate.* Light blue, cryst powder. Slightly soluble in cold water, more in hot water; sol in alc or in dil acids with separation of benzoic acid.

Lead salt dihydrate. $C_{14}H_{10}O_4Pb.2H_2O$, *lead benzoate.* Cryst powder. *Poisonous!* Slightly sol in water.

Manganese salt tetrahydrate, $C_{14}H_{10}MnO_4.4H_2O$, *manganese benzoate.* Pale-red powder. Sol in water, alc. Also occurs with $3H_2O$.

Nickel salt trihydrate, $C_{14}H_{10}NiO_4.3H_2O$, *nickel benzoate.* Light-green odorless powder. Slightly sol in water; sol in ammonia; dec by acids.

Potassium salt trihydrate, $C_7H_5KO_2.3H_2O$, *potassium benzoate.* Cryst powder. Sol in water, alc.

Silver salt. $C_7H_5AgO_2$, *silver benzoate.* Light-sensitive powder. Sol in 385 parts cold water, more sol in hot water; very slightly sol in alc.

Uranium salt, $C_{14}H_{10}O_6U$, *uranium benzoate, uranyl benzoate.* Yellow powder. Slightly sol in water, alc.

Toxicity: Mild irritant to skin, eyes, mucous membranes.

USE: Preserving foods, fats, fruit juices, alkaloidal solns, etc: manuf benzoates and benzoyl compds, dyes; as a mordant in calico printing: for curing tobacco. As standard in volumetric and calorimetric analysis.

THERAP CAT: Pharmaceutic aid (antifungal agent).

THERAP CAT (VET): Has been used with salicylic acid as a topical antifungal.

the first entry on page 216. The actual page number is at the bottom left of the page.

2. ***n-Butyl Bromide.*** Listed as a substituted butane with the systematic name given as a synonym.

3. ***C 35.06%,*** ... Elemental analysis data; the percent of each element in the compound.

4. ***Prepd from.*** ... A short note on how 1-bromobutane has been prepared, and references to the original literature (journals).

5.   *1.2686.* The tiny 25 over 4 makes this a specific gravity. Note that the temperatures are given with the **d** and not with the numerical value as in *Lange's* and the *CRC*.

Entry: Benzoic Acid (Fig. 15)

1. *Line 2. dracylic acid.* What a synonym! Label your benzoic acid bottles this way and no one will ever "borrow" your benzoic acid again.
2. *Lines 3–7.* Natural sources of benzoic acid.
3. *Lines 7–9.* Industrial syntheses of benzoic acid. These are usually not appropriate for your lab bench preparations.
4. *Lines 9–20.* References to the preparation and characteristics of benzoic acid in the original literature (journals).
5. *Structure.* A structural formula of benzoic acid.
6. *Lines 21–40. Physical data.* The usual crystalline shape, density (note *two* values reported.), sublimation notation, boiling point data, and so on. K at 25° is the ionization constant of the acid; the pH of the saturated solution (2.8 at 25°C) is given. The solubility data *(Soly)* are very complete, including water solutions at various temperatures, a bit about the phase diagram of the compound, and solubility in other solvents. Note that numerical data are given where possible.
7. *Lines 41–67.* Properties of some slots of benzoic acid.
8. *Line 68.* Toxicity data for benzoic acid.
9. *Lines 69–72.* Some commercial uses of benzoic acid.
10. *Lines 73–75.* Therapeutic uses, both human and veterinary, for benzoic acid.

If all the chemical entries were as extensive as the one for benzoic acid, this would be the handbook of choice. Because benzoic acid has wide use in medicine and food production, and it is very important to know the physical properties of drugs and food additives, a lot of information on benzoic acid winds up in the *Index*. 1-Bromobutane has little such use, and the size of the entry reflects this. Unfortunately, many of the compounds you come in contact with in the organic laboratory are going to be listed with about the same amount of information you'd find for 1-bromobutane, and not with the large quantities of data you'd find with benzoic acid.

THE ALDRICH CATALOG

(**The Aldrich Catalog.** *Aldrich Chemical Co., Inc., Milwaukee, Wisconsin.*) Not your traditional hard-bound reference handbook, but a handy book nonetheless. The company makes many compounds, some not yet listed in the other handbooks, and often gives structures and physical constants for them. Because Aldrich is in the business of selling chemicals to industry, many industrial references are given.

3

Entry: 1-Bromobutane (Fig. 16)

1. ***1-Bromobutane.*** Here it is listed *strictly* alphabetically as it is—with all the bromo-compounds—not as a butane, 1-bromo-, and only a cross reference as a butyl bromide.
2. ***[109-65-9].*** This is the Chemical Abstracts Service (CAS) Registry number. *Chemical Abstracts*, published by the American Chemical Society, is a listing of the abstract or summary written for any paper in the chemical literature. Every compound made gets a number. This makes for easy searching by computer as well as by hand.
3. ***bp 100–104°.*** Without a tiny superscript this is the boiling point at 760 torr.
4. ***n_D^{20} 1.4390.*** Index of refraction. The temperature (20°) modifies the n, rather than the number as in the *CRC*.

	Benzoic acid, 99+ %, GOLD LABEL, A.C.S. reagent	**500g**	**17.20**
	[05-05-0]	**3kg†**	**80.65**
24,238-1	$C_6H_5CO_2H$ FW 122.12 mp 122-123° bp 249°		
★	Fp 250°F(121°C) *Beil.* **9**,92 *Fieser* **1**,49 *Merck*		
	Index **10**,1093 *FT-IR* **1**(2),186A *MSD Book* **1**,160A		
	RTECS# DG0875000 Disp. A *IRRITANT*		
10,947-9	**Benzoic acid,** 99%, [65-85-0]	**500g†**	**7.00**
★	$C_6H_5CO_2H$	**3kg†**	**31.00**
23,988-7	**1-Bromobutane,** 99+ %, GOLD LABEL [*109-65-9*]	**50g**	**15.75**
★	(*n*-butyl bromide)		
	$CH_3(CH_2)_3Br$ FW 137.05 mp −112° bp 100-104°		
	n_D^{20} 1.4390 d 1.276 Fp 75°F(23°C) *Beil.* **1**,119 *Merck*		
	Index **10**,1526 *MSD Book* **1**,236B *RTECS#* EJ6225000		
	Disp. D *FLAMMABLE LIQUID IRRITANT*		
B5,949-7	**1-Bromobutane,** 99%, [*109-65-9*] (*n*-butyl bromide)	**500g**	**17.40**
★	$CH_3(CH_2)_3Br$	**1kg**	**23.10**

Fig. 16 Sample entries from the *Aldrich Catalog*, 1986–87.

5. **_d 1.276._** The density in g/cc.
6. **_Fp 75°F(23°C) Flash Point._** Above 75°F, a mixture of 1-bromobutane and air and a spark will go up like gangbusters. Watch out!
7. **_Beil. 1, 119._** The Beilstein reference; Volume 1, page 119.
8. **_Merck Index 10, 1526._** The _Merck Index,_ 10th ed. reference; compound #1526 (Fig. 15).
9. **_MAS Book 1, 236B._** A reference to the page location of the entry in the _Sigma-Aldrich Library of Chemical Safety Data_, Edition 1.
10. **_RTECS# EJ6225000._** The reference number in the _Registry of Toxic Effects of Chemical Substances (RTECS)._ 1-Bromobutane is on the inventory of the Environmental Protection Agency (EPA) according to the Toxic Substances Control Act, PL9469, October 11, 1976 (TSCA).
11. **_Disp D._** There are methods of disposal given in the _Aldrich Catalog._ Go to method D and throw 1-bromobutane out according to the rules. Remember, the methods given are for the disposal of large amounts of a single substance, as might be found in an industrial application. The rules for the disposal of the waste generated in your undergraduate laboratory may differ considerably.
12. **_Flammable liquid irritant._** Yep, it sure is.

Note the differences in prices for the 99+% GOLD LABEL and the merely 99% 1-bromobutane. Before you buy, check on the use of the chemical. Normally, you can buy the least expensive grade of the chemical, and distill or recrystallize it yourself before you use it, if necessary.

Entry: Benzoic Acid (Fig. 16)

1. **_Fieser 1, 49._** A reference to Fieser & Fieser's _Reagents for Organic Synthesis_, Volume 1, page 49. This multivolume series gives the syntheses and reactions of many organic compounds, along with references to the original literature.
2. **†.** Benzoic acid cannot be shipped by parcel post.
3. **Beil. 9, 92.** A reference to Beilstein, Volume 9, page 92.
4. **_FT-IR 1(2)186A._** The Fourier-Transform Infra-Red spectrum of benzoic acid is in Edition 1, Volume 2, page 186A of _The Aldrich Library of FT-IR Spectra._

NOT CLEAR—CLEAR?

One antonym for **clear** is **cloudy**. Another antonym for **clear** is **colored**. When you say you "obtained a clear liquid," do you mean that it is not cloudy or that it is colorless?

Cloudiness usually means you've gotten water in your organic liquid. Colorless should be self-explanatory. You should always pair the turbidity and color designations:

"a clear, colorless liquid"
"a clear, blue liquid"
"a cloudy, colorless liquid"
"a cloudy, blue liquid"

I use clear to mean not cloudy, and *water-white* to mean not colored. Water-white is a designation found in the older chemical literature; **colorless** is more modern.

Is that clear?

GENERIC
JOINTWARE

4

Using **standard taper jointware,** you can connect glassware without rubber stoppers, corks, or tubing. Pieces are joined by glass connections built into the apparatus (Fig. 17). They are manufactured in standard sizes, and you'll probably use ⑂ 19/22.

The symbol ⑂ means **standard taper**. The first number is the size of the joint at the widest point, in millimeters. The second number is the length of the joint, in millimeters. This is simple enough. Unfortunately, life is not all that simple, except for the mind that thought up this next devious little trick.

STOPPERS WITH ONLY ONE NUMBER

Sounds crazy, no? But with a very little imagination, and even less thought, grave problems can arise from confusing the two. Look at Fig. 18, which shows all glass stoppers are not alike. Interchanging these two leads to **leaking joints** through which your **graded** product can escape. Also, the ⑂ 19/22 stopper is much more expensive than the ⑂ 19 stopper, and you may *have to pay money* to get the correct one when you check out at the end of the course. Please note the emphasis in those last two sentences. I appeal to your better nature and common sense. So, take some time to check these things out.

As you can see from Fig. 18, that single number is the width of the stopper at its top. There is no mention of the length, and you can see that it is too short. The ⑂ 19 stopper *does not* fit the ⑂ 19/22 joint. Only the ⑂ 19/22 stopper can fit the ⑂ 19/22 joint. Single-number stoppers are commonly

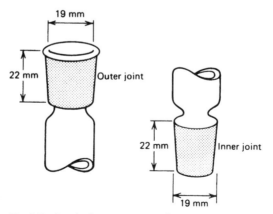

Fig. 17　Standard taper joints (⑂19/22).

used with volumetric flasks. Again, they will leak or stick if you put them in a double-number joint.

With these delightful words of warning, we continue the saga of coping with ground-glass jointware. Figure 19 shows some of the more familiar pieces of jointware you may encounter in your travels. They may not be so familiar to you now, but give it time. After a semester or so, you'll be good friends, go to reactions together, maybe take in a good synthesis. Real fun stuff!

These pieces of jointware are the more common pieces that I've seen used in the laboratory. You may or may not have *all* the pieces shown in Fig. 19. Nor will they necessarily be called exactly by the names given here. The point is *find out* what each piece is, and *make sure* that it is in good condition *before* you sign your life away for it.

ANOTHER EPISODE OF LOVE OF LABORATORY

"And that's $28.46 you owe us for the separatory funnel."
"But it was broken when I got it!"
"Should've reported it then."
"The guy at the next bench said it was only a two-dollar powder funnel and not to worry and the line at the stockroom was long anyway, and . . . and . . . anyway the stem was only cracked a little . . . and it worked O.K. all year long. . . . Nobody said anything. . . ."
"Sorry."

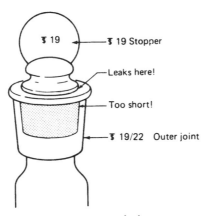

Fig. 18   A ℥29 nonstandard stopper in a ℥19/22 standard taper joint.

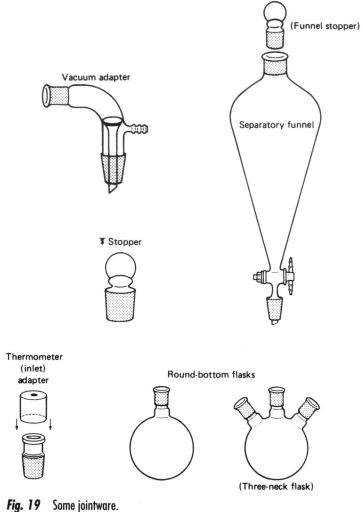

Fig. 19 Some jointware.

Tales like these are commonplace, and ignorance is no excuse. Don't rely on expert testimony from the person at the next bench. He may be more confused than you are. And equipment that is "slightly cracked" is much like a person who is "slightly dead." There is no in-between. If you are told that you *must* work with damaged equipment because there is no replacement available, you would do well to get it in writing.

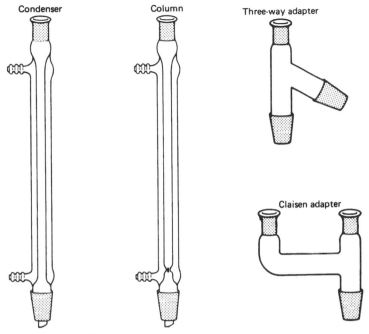

Fig. 19 (continued)   Some jointware.

HALL OF BLUNDERS AND THINGS NOT QUITE RIGHT

Round-Bottom Flasks

Round-bottom (R.B.) jointware flasks are so round and innocent looking that you would never suspect they can turn on you in an instant.

1. ***Star cracks.*** A little-talked-about phenomenon that turns an ordinary R.B. flask into a potentially explosive monster. Stress, whether prolonged heating in one spot or indiscriminate trouncing upon hard surfaces, can cause a flask to develop a **star crack** (Fig. 20) on its backside. Sometimes they are hard to see, but if overlooked, the flask can split asunder at the next lab.

2. ***Heating a flask.*** Since they are cold-blooded creatures, flasks show more of their unusual behavior while being heated. The behavior is usually unpleasant if certain precautions are not taken. In addition to star cracks, various states of disrepair can occur, leaving you with a benchtop to clean. Both humane and cruel heat treatment of flasks

Fig. 20 R.B. flask with
star crack.

will be covered in Chapter 18, "Sources of Heat," which is on the SPCG (Society for the Prevention of Cruelty to Glassware) recommended readings list.

Columns and Condensers

A word about **distilling columns** and **condensers:**

Different!

Use the **condenser** as is for **distillation** and **reflux** (see Chapter 20, "Distillation," and Chapter 22, "Reflux"). You can use the *column with or without column packing* (bits of metal or glass or ceramic or stainless-steel sponge—whatever)! That's why the column is wider and it has *projections* at the end (Fig. 21). These projections help hold up the column packing if you use any packing at all (see Fig. 104).

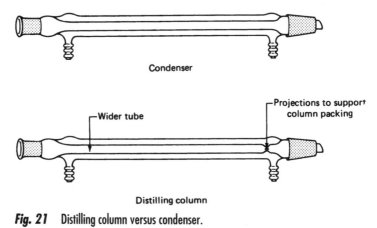

Fig. 21 Distilling column versus condenser.

If you jam column packing into the skinny condenser, the packing may never come out again! Using a condenser for a packed column is bad form and can lower your esteem or grade, whichever comes first.

You might use the column as a condenser.
Never use the condenser as a packed column!

The Adapter with Lots of Names

Figure 22 shows the one place where joint and nonjoint apparatus meet. There are two parts: a rubber cap with a hole in it and a glass body. Think of the rubber cap as a rubber stopper through which you can insert thermometers, inlet adapters, drying tubes, and so on.

CAUTION! Do not force. You might snap the part you're trying to insert. Handle both pieces through a cloth; lubricate (water) and then insert carefully.

The rubber cap fits over the **nonjoint** end of the glass body. The other end is a **ground glass joint** and *fits only other glass joints*. The rubber cap should neither crumble in your hands nor need a 10-ton press to bend it. If the cap is shot, get a new one. Let's have none of these corks, rubber stoppers, chewing gum, or any other type of plain vanilla adapter you may have hiding in the drawer.

And remember: Not only thermometers, but **anything** that resembles a glass tube can fit in here! This includes unlikely items such as **drying tubes** (they have an outlet tube) and even a **funnel stem** (you may have to couple the stem to a smaller glass tube if the stem is too fat).

The imaginative arrangements shown in Fig. 23 are acceptable.

Forgetting the Glass

Look, the Corning people went to a lot of trouble to turn out a piece of glass (Fig. 24) that fits perfectly in *both* a glass joint *and* a rubber adapter, *so use it!*

Inserting Adapter Upside Down

This one (Fig. 25) is really ingenious. If you're tempted in this direction, go sit in the corner and repeat over and over,

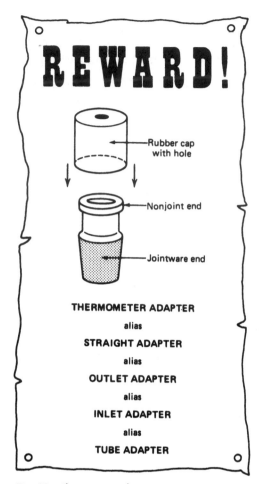

Fig. 22 Thermometer adapter.

"Only glass joints fit glass joints."

Inserting Adapter Upside Down *Sans* Glass

I don't know whether to relate this problem (Fig. 26) to glass forgetting, or upside-downness, since both are involved. Help me out. If I don't see you trying to use an adapter upside down without the glass, I won't have to make such a decision. So, don't do it.

GREASING THE JOINTS

In all my time as an instructor, I've never had my students go overboard on greasing the joints, and they never got them stuck. Just lucky, I guess. Some instructors, however, use grease with a passion and raise the roof over it. The entire concept of greasing joints is not as slippery as it may seem.

To Grease or Not to Grease

Generally, you'll grease joints on two occasions. One, when doing vacuum work to make a tight seal that can be undone; the other, doing reactions with a strong base that can etch the joints. Normally, you don't have to protect the joints during acid or neutral reactions.

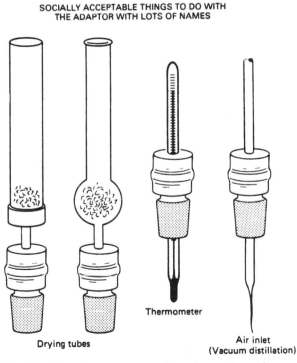

SOCIALLY ACCEPTABLE THINGS TO DO WITH
THE ADAPTOR WITH LOTS OF NAMES

Thermometer

Drying tubes

Air inlet
(Vacuum distillation)

Fig. 23 Unusual, yet proper, uses of the adapter with lots of names.

Fig. 24 The glassless glass adapter.

Preparation of the Joints

Chances are you've inherited a set of jointware coated with 47 semesters of grease. First wipe off any grease with a towel. Then soak a rag in any hydrocarbon solvent (hexane, ligroin, petroleum ether—and *no flames*, these burn like gasoline) and wipe the joint again. Wash off any remaining grease with a strong soap solution. You may have to repeat the hydrocarbon-soap treatments to get a clean, grease-free joint.

Fig. 25 The adapter stands on its head.

Into the Grease Pit (Fig. 27)

First, use only enough grease to do the job! Spread it thinly along the upper part of the joints, only. Push the joints together with a twisting motion. The joint should turn clear from one third to one half of the way down the joint. *At no time should the entire joint clear!* This means you have *too much grease* and you must start back at *Preparation of the Joints*.

Don't interrupt the clear band around the joint. This is called **uneven greasing** and will cause you headaches later on.

STORING STUFF AND STICKING STOPPERS

At the end of a grueling lab session, you're naturally anxious to leave. The reaction mixture is sitting in the joint flask, all through reacting for the day, waiting in anticipation for the next lab. You put the correct glass stopper in the flask, clean up, and leave.

The next time, the stopper is stuck!

Stuck but good! And you can probably kiss your flask, stopper, product, and grade goodbye!

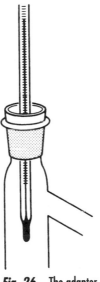

Fig. 26 The adapter on its head, without the head.

Frozen!

Some material has gotten into the glass joint seal, dried out, and cemented the flask shut. There are few good cures but several excellent preventive medicines.

Corks!

Yes, corks, old-fashioned, non-stick-in-the-joint corks.

If the material you have to store *does not attack cork*, this is the cheapest, cleanest method of closing off a flask.

A well-greased glass stopper *can* be used for materials that attack cork, but *only* if the stopper has a good coating of stopcock grease. Unfortunately, this grease can get into your product.

Do not use rubber stoppers!

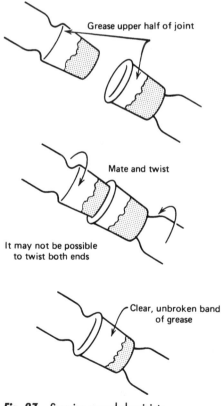

Grease upper half of joint

Mate and twist

It may not be possible
to twist both ends

Clear, unbroken band
of grease

Fig. 27 Greasing ground glass joints.

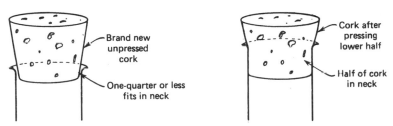

Fig. 28 Corking a vessel.

Organic liquids can make rubber stoppers swell up like beach balls. The rubber dissolves and ruins your product, and the stopper won't come out either. Ever.

The point is

Dismantle all ground glass joints before you leave!

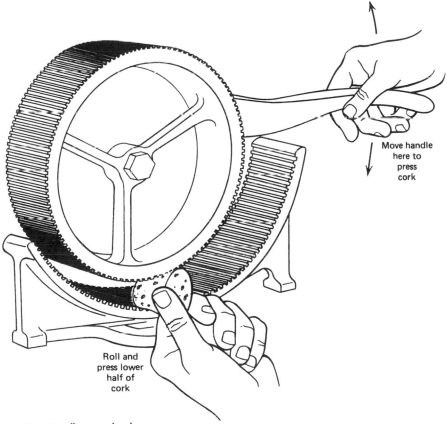

Fig. 29 A wall-mounted cork press.

CORKING A VESSEL

If winemakers corked their bottles like some people cork their flasks, there'd be few oenophiles and we'd probably judge good years for salad dressings rather than wines. You don't just take a new cork and stick it down into the neck of the flask, vial, or what have you. You must press the cork first. Then as it expands, it makes a very good seal and doesn't pop off.

A brand new cork, **before pressing** or **rolling**, should fit only about one quarter of the way into the neck of the flask or vial. Then you roll the lower half of the cork on your *clean* benchtop to soften and press the small end. *Now* stopper your container. The cork will slowly expand a bit and make a very tight seal (Fig. 28).

THE CORK PRESS

Rather than rolling the cork on the benchtop, you might have the use of a **cork press**. You put the small end of the cork into the curved jaws of the press, and when you push the lever up and down, the grooved wheel rolls and mashes the cork at the same time (Fig. 29). Mind your fingers!

OTHER INTERESTING EQUIPMENT

An early edition of this book illustrated some equipment specific to the State University of New York at Buffalo, since that's where I wrote it. It's now a few years later, and I realize that you can't make a comprehensive list.

Buffalo has an unusual pear-shaped distilling flask that I've not seen elsewhere. The University of Connecticut equipment list contains a Bobbitt Filter Clip that few other schools have picked up.

So if you are disappointed that I don't have a list and drawing of every single piece of equipment in your drawer, I apologize. Only the most common organic lab equipment is covered here. Ask your instructor "Whattizzit?" if you do not know.

I assume that you remember Erlenmeyer flasks, and beakers and such from the freshman lab. I'll discuss the other apparatus as it comes up in the various techniques. This might force you to read this book before you start lab.

Check out Fig. 30. Not all the mysterious doodads in your laboratory drawer are shown, but the more important ones are.

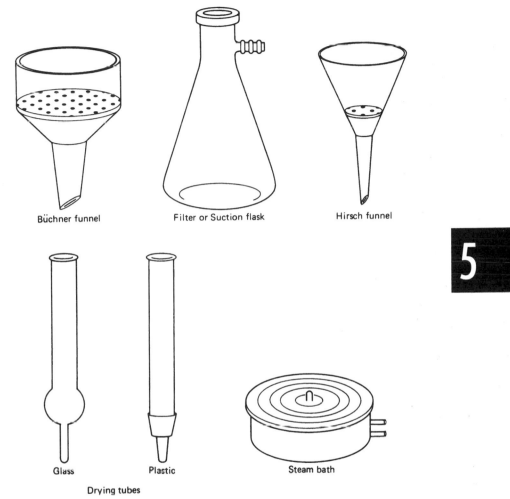

Büchner funnel   Filter or Suction flask   Hirsch funnel

Glass   Plastic   Steam bath

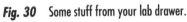

Drying tubes

Fig. 30 Some stuff from your lab drawer.

MICROSCALE
JOINTWARE

6

MICROSCALE: A FEW WORDS

Recently, **microscale** laboratory equipment has appeared in the under-graduate laboratory and has materially changed some of the operations and equipment you'll be using. Pay attention:

1. *Microscale is small.* How small is it? Well there's no absolute cutoff, but some suggest 10 g is macroscale, 1 g is semi-microscale, and 0.1 g is truly microscale. Now 0.1 g of a **solid** is really something. A KBr (potassium bromide) infrared pellet uses 0.01 g (10 mg) and an NMR (nuclear magnetic resonance), taking five times as much (50 mg), leaves you with 40 mg to play with (melting point, chemical tests, hand in, and so on). Now 0.1 g of a **liquid** is really nothing. Liquids are difficult to divide; they, well, *flow.* And if you assume there're about 20 drops per milliliter for the average liquid, and the average liquid has a density of 1 g/ml, you've got two whole microscale drops. One drop for the IR, one for the NMR and for the boiling point... whoops!

   Fortunately, most experimenters bend the scale so that you get close to 1g (approximately 1 ml) when you prepare a liquid product. *That does not give you license to be sloppy!* Just be careful. Don't automatically put your liquid product in the biggest container you have. A *lot* of liquid can lose itself very quickly.

2. *Microscale is new.* Not brand-spanking-new, but new enough that there's no real standardization on equipment. Or should I say there are several standards. I'm going with tradition, and sticking largely to the kinds of things described by Mayo, Pike, and Butcher in *Microscale Organic Laboratory.* These guys use real equipment with real ground-glass joints that often look vaguely like organic chemistry laboratory setups. Some just carry out lots of reactions in very long test tubes ("reaction tubes"), which is fine but not nearly as much fun.

I've put drawings of microscale equipment I've had occasion to use in this section, along with some discussion of the O-ring seals, conical vials, drying tubes, and so on. I've put full descriptions of certain microscale apparatus with the operations they're used in. So Craig tubes show up with recrystallization; the Hickman still is with distillation.

THE "O-RING CAP SEAL"

In the old days, you would put two glass joints together and, if you wanted a vacuum tight seal, you'd use a little grease. These days, you're told to use a plastic cap and and an O-ring to get a greaseless, high-tech seal. Maybe. Not all O-ring cap seal joints are created equal.

In a famous microscale manual, you're told how to make one of these seals between an air condenser and a conical vial from the top down. In sum:

1. Get a conical vial (appropriate size) that'll fit on the bottom end (male joint) of the wide-bottom air condenser (Fig. 31).
2. Drop the O-ring over the top of this air condenser.
3. Drop the cap over the top of this air condenser.
4. Carefully screw the cap down onto the threads of the vial.

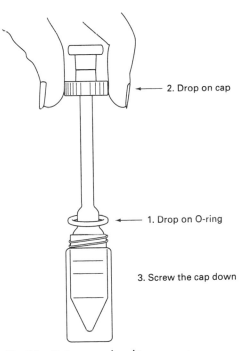

Fig. 31   O-ring cap seal on skinny apparatus.

Unfortunately, you *can't* drop an O-ring and cap over a water-jacketed condenser, Hickman still, and so on. What to do? Using the water-jacketed condenser as an example, go from the bottom up:

1. Get a conical vial (appropriate size) that'll fit on the end (male joint) of the condenser (Fig. 32).
2. Take the vial **off** the end.
3. Put a plastic cap onto the male joint and hold it there.
4. Push an O-ring up over the male joint onto the clear glass. The O-ring *should* hold the cap up. (Fingers might be necessary.)
5. Put the conical vial *back* on the male joint.
6. Carefully screw the cap down onto the threads of the vial.

Sizing Up the Situation

As I said, microscale equipment design is in a state of flux, but there are a few things to note about the O-ring cap seal:

1. ***A 3 to 5 ml conical vial and ⊤ 14/10 joints.*** The plastic caps for these sizes have holes cut in them that are *extremely* tight. These

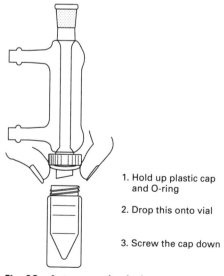

1. Hold up plastic cap and O-ring

2. Drop this onto vial

3. Screw the cap down

Fig. 32 O-ring cap seal on husky apparatus.

make a really good seal, and I suspect you would break the glass before *ever* pulling one of these babies apart. Once, when dismantling one of these joints, I caught the O-ring in the cap and as I pulled the cap off the joint, the cap sliced the O-ring in two. Watch it.

2. *A 0.3 to 1.0 ml conical vial and ͳ 7/10 joint.* (Fig. 33) The ͳ joint rattles around in the hole in the plastic caps for this size. As a consequence, whenever you tighten these joints, the O-ring squeezes out from under the cap. Yes, the joint's tight, but gas-tight? You can pull the joints apart with modest effort, and, if the O-ring has squeezed out entirely, forget the cap as a stabilizing force.

3. *A 0.1 ml conical vial and ͳ 5/5 joint.* For these sizes, the hole in the cap is again snug around the male joint so that when you screw down the cap, the O-ring doesn't squeeze out. This joint also appears to be very tight.

Why I Don't Really Know How Vacuum Tight These Seals Are

I'll take a break here, and let Kenneth L. Williamson of *Macroscale and Microscale Organic Experiments* (D. C. Heath & Co., 1989), p. 102, take over. Here he is on the subject of Microscale Vacuum Distillation Assemblies: "On a truly micro scale (<10mg) simple distillation is not practical because of mechanical losses."

That's clear enough. You lose lots of your microscale product with a reduced-pressure (vacuum) distillation. As far as I know, no microscale laboratory manual has anyone perform this. Yes, you might be asked to

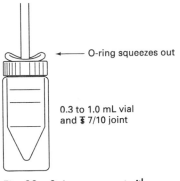

O-ring squeezes out

0.3 to 1.0 mL vial
and ͳ 7/10 joint

Fig. 33 O-ring squeeze-out with ͳ 7/10 joint.

remove solvent under reduced pressure, but that's neither a reduced-pressure distillation nor uniquely microscale.

So, since nobody actually has any microscale reduced-pressure distillations, and with certain sizes of equipment the O-rings can squeeze out from under the plastic caps, and since I don't use them for reduced-pressure (vacuum) distillation anyway (I switch to a semi-micro apparatus with real 14/20 joints and—gasp!—grease), I don't know how vacuum tight these seals really are.

THE COMICAL VIAL (THAT'S *CONICAL!*)

The **conical vial** (Fig. 34) is the round-bottom flask of the microscale set, with considerably more hardware. When your glassware kit is brand new, every vial has a plastic cap, O-ring, and plastic disk to match. After one semester, every vial *needs* this stuff. When you check-in, make sure that every conical vial has a cap that fits and that you have at least one O-ring for each and a plastic disk that just fits inside this cap.

The Conical Vial As Vial

You would think that with an O-ring and a Teflon-faced plastic disk wedged under the cap, this would make a very tight seal. Think again. One of my

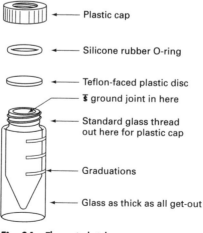

Plastic cap

Silicone rubber O-ring

Teflon-faced plastic disc

$\mathbf{\mathsf{T}}$ ground joint in here

Standard glass thread out here for plastic cap

Graduations

Glass as thick as all get-out

Fig. 34 The conical vial.

students made n-hexyl amine as a special project and packaged it in one of these vials; some of the amine leaked out, air got in, and the amine oxidized to a lovely brown color. You can still use these if you want but be careful:

1. Check to see that the Teflon-faced plastic disk doesn't have any holes in it. Sometimes you get to pierce a disk with a syringe needle in order to add something to a setup. You use the disk as a puncturable seal. With a puncture or two, the seal won't be as good as with a fresh disk. Just to be sure, get a new one if your disk has had lots of punctures.
2. The Teflon side to the chemicals, please. Especially if the compound might attack the plastic disk. Teflon is *inert*; the plastic is *ert*.
3. The O-ring goes close to the cap. Otherwise, it can't press the disk down to the glass.
4. Now the cap. Screw this baby down and you should, *should,* mind you, have a very well-sealed vial.

Laboratory Gorillas Please Note
Overtightening Caps Breaks the Threads
Thank You

Even so, watch for leaks! Sometimes you're asked to perform an extraction in one of these vials, and you decide to shake the bejesus out of the vial to get the two layers to mix. (See Extraction and Washing: Microscale.) After you clean your fingers, you *will* be more gentle when you repeat the experiment, won't you?

Packaging Oops

Did you get the weight of the product so that you could calculate the yield? ("What? *Calculation* in organic lab?") Fat lot of good that tight seal will do for you now that you have to undo it to weigh your product. And you'll lose a lot of your product on the walls of the vial. Nothing but problems. Especially in microscale.

Tare to the Analytical Balance

You can avoid losing your product and save a bit of time if you weigh the empty vial *before* you put your product in it. Microscale quantities, unfortunately, may require a high-precision **analytical balance** rather than a **top-loading balance**, and you should be aware of at least one thing:

Analytical balances are very delicate creatures.
Only closed containers in the balance box!

Now I don't know if you have a hanging pan analytical balance with dialup weights or a fancy electronic model, and I don't care. Just keep your noxious organic products in closed containers, OK?

Electronic Analytical Balance

1. Open your notebook and, on one line, write "Mass of empty container," and get ready to write the mass in. What? You didn't bring your notebook with you to record the weights? You were going to write the numbers down on an old paper towel? *You fail.* Go back and get your notebook.
2. Turn the balance on. If it's already on, zero the balance. See your instructor for the details.
3. Open the balance case door, and put the empty vial with cap and O-ring and plastic disk on the balance pan. Close the balance case door.
4. Wait for the number to stop jumping around. Even so, electronic digital balances can have a "bobble" of one digit in the least significant place. Write down the most stable number.
5. Take the vial out of the balance case. *Now* fill the vial with your product. Close it up. No open containers in the balance case.
6. Write this new number down. Subtract the numbers. You have the weight of your product, with the minimum of loss.

"What if I have a **Tare bar** or **Tare knob**? Why not use this to zero the balance before putting in product. That way, I don't have to subtract—just get the weight of the product directly." Sure. Now take some of your product out of the vial to use in, say, another reaction. How much did you take? What? You say you already put it into the other reaction vial, and it's all set up and you can't take product out to weigh it? What? You say you think you'll lose too much product re-weighing it in yet another vial before you commit it to your reaction? Did you record the tare of the product vial? No? One subtraction too many for you, huh?

But if you did record the weight of the empty vial, cap, and O-ring, all you need do is remove some product, close the vial up, weigh it again, and subtract **this** weight from the sum of the tare and original product mass.

"What if I have a top-loading balance without a cage—do I have to have only closed containers on it while weighing?" Yes. Cuts down on the mess. An open container on a balance pan is usually an irresistible invitation to load product into it while on the pan, spilling the product onto the pan.

"What about weighing paper?" What *about* weighing paper. You *always* fill or empty your containers **away from the balance pan**, and you always weigh your samples *in these closed containers,* so what about weighing papers?

Heating These Vials

Sources of heat vary according to what you have available. Some use a can-type heating mantle, only it's filled with sand. You raise or lower the vial in the sand pile to set the temperature you want. Others use a **crystallization dish**, on top of a **hot plate with magnetic stirrer** (Fig. 35). You put the vial flat on the crystallization dish and add sand. You can monitor the temperature of the vial in a rough fashion by sticking a thermometer in the sand.

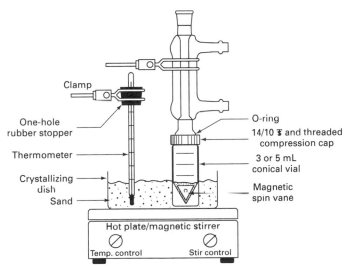

Fig. 35 Hot bath for conical vial; just happens to be a reflux.

THE MICROSCALE DRYING TUBE

The **microscale drying tube** (Fig. 36) is just a right-angled glass tube with a ᵀ joint on one end. You put a small cotton plug in first, and then load in the drying agent. Finish up with another cotton plug.

If you let these sit around too long, the drying agent cakes up and you could break the tube hacking this stuff out. It's actually cheaper to empty the tube after using it. If you can't hack that, soak the tube overnight (or so) and dissolve the caked drying agent out.

GAS COLLECTION APPARATUS

Several microscale kits have a **capillary gas delivery tube**. With this tube attached, you can collect gaseous products.

1. Put a cap on your **gas collection reservoir**. You can calibrate the tube by adding 2, 3, or 4 ml of water and marking the tube. Some have calibration lines. Check it out (Fig. 37).
2. Get a 150 ml beaker and fill it with water to about 130 ml. A 100 ml beaker filled to 80 ml will work, but it'll be a bit tight.
3. Now fill the **gas collection reservoir** with water, put a finger over the open end, invert the tube, and stick the tube (and your fingers) under the water. Remove your finger. See that the gas collection reservoir remains filled.
4. Angle the straight end of the **capillary gas delivery tube** under and into the gas collection reservoir (Fig. 38). Believe it or not, the gas

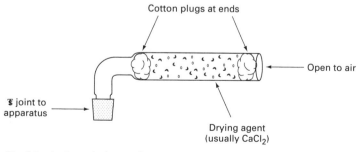

Fig. 36 A microscale drying tube.

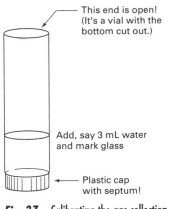

This end is open!
(It's a vial with the
bottom cut out.)

Add, say 3 mL water
and mark glass

Plastic cap
with septum!

Fig. 37 Calibrating the gas collection
reservoir.

collection reservoir, sitting atop the capillary gas delivery tube, is
fairly stable, so you can put this aside until you need it.

5. When you're ready, use a ring with a screen to support the beaker at
the right height necessary to connect the joint end of the capillary gas
delivery tube to whatever you need to. Direct connection to a conical
vial usually uses an O-ring cap seal; connection to a condenser may not
(Fig. 39). Note that I've drawn the water connections to the condenser
at an angle. Although the drawing is two-dimensional, your lab bench
is not. You may have to experiment to find an arrangement that will
accommodate the ringstand, hot plate, and other paraphernalia.
Don't get locked into linear thinking.

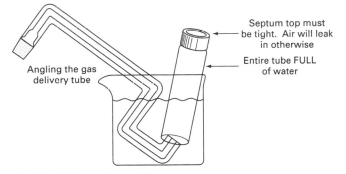

Septum top must
be tight. Air will leak
in otherwise

Entire tube FULL
of water

Angling the gas
delivery tube

Fig. 38 Setting up for gas collection.

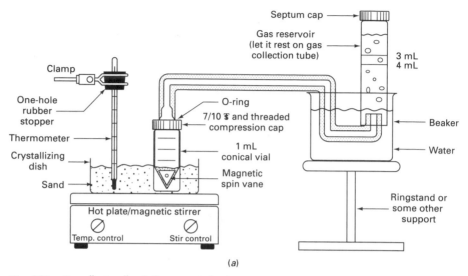

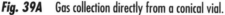

Fig. 39A   Gas collection directly from a conical vial.

Generating the Gas

Usually, you warm the conical vial somehow to have the reaction generate the gas you want to collect.

Do not stop heating once you start!
**Water will get sucked back into the reaction mixture
as the apparatus cools!**

Watch the flow of gas into the collection bottle. If you increase the heat and no more gas comes out, the reaction is probably done.

Remove the capillary gas delivery tube first.
Then remove the heat.

Of course, some gaseous reaction product, a few microliters perhaps, will still come off the mixture. Don't breathe this. And you *did* carry this out in a hood, no?

If your capillary gas delivery tube was connected to the top of a condenser (Fig. 39b), just pick the beaker, collection vial, and capillary gas delivery tube up, as a unit, off of the condenser and set it down. Then attend to the reaction.

If your capillary gas delivery tube was connected directly to a conical vial with an O-ring cap seal (Fig. 39a), you have to undo the seal on a hot vial (!)

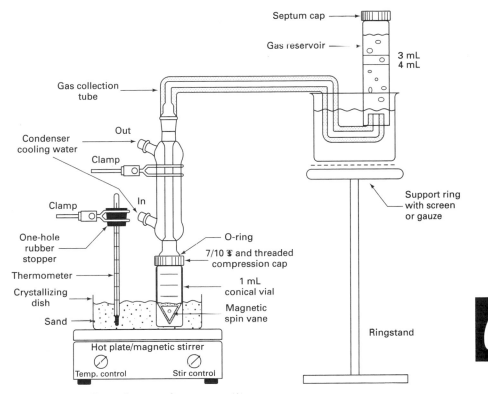

Fig. 39B Gas collection from a condenser.

in order to lift the entire system off as a unit. And you can't let the vial cool with the capillary gas delivery tube under water, or you'll get water sucked back into the vial. The best I've been able to do is carefully lift the gas collection vial off the gas delivery tube (Watch the water level in the beaker. If it's too low, the water in the collection vial will run out!) and then lower the beaker with the collection vial in it away from the capillary gas delivery tube. There's got to be a better way, but I don't know it yet. Ask your instructor.

Isolating the Product

With the **gas collection reservoir**, you have an O-ring cap seal with a Teflon-faced plastic insert. Pierce the insert with a hypodermic syringe and draw off some of the product. Usually, you put this gas through GC (see Chapter 30, "Gas Chromatography," for particulars).

Some Present-Day Gas-Collecting Ironies

Life in the organic laboratory does not usually present us with situations that have a modicum of wry humor, and when they occur they must be cherished. For instance none of the microscale kits sold by Ace Glass—the pioneers in this field—or its competitor, Corning Glass Works, includes the **gas collection reservoir**. It is sold separately, much as batteries are. Ace Glass Bulletin 898 has this beast in "typical undergraduate setups" on the cover, page 3, and page 5 in order to sell their microscale kits, but it's not actually *in* any of their kits. *Caveat emptor.*

The usual capillary gas delivery tube has a ⴲ 7/10 joint—and we've seen that this size makes the least gas-tight seal. So, of course, it's on the gas delivery tube. And some condensers don't even *have* a thread on their ⴲ 7/10 outlet to tighten the cap on anyway. So much for greaseless gas-tight seals.

Finally, although I experimented a bit with this gas collection setup, mostly I read up on the technique in Mayo et al. By now they should have another edition out, but check the photo on the cover of the 1985 work: There it is—the entire gas collection setup in full-color glory—with the end of a Craig tube sitting on the capillary gas collection tube. I guess the batteries are extra even if you're the group that designed the batteries.

PIPET TIPS

The Pasteur pipet (Fig. 40) is really handy for *all* scales of laboratory, not just microscale, but it's in microscale that it gets the most use. It usually comes in two sizes, a 9 inch, approximately 3 ml pipet, and a 5 ¾ inch, approximately 2 ml pipet.

PRE-PREPARING PASTEUR PIPETS

Figure 40 shows a rough calibration chart for both sizes.

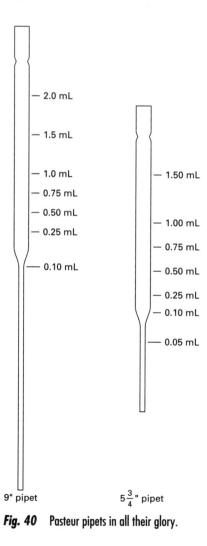

Fig. 40 Pasteur pipets in all their glory.

Calibration

You should get an idea about how high in the pipet 0.5, 1.0, 1.5, and 2.0 ml of liquid are. Just draw measured amounts into the tube and look. I'd mark **one** pipet with all these volumes and keep it as a reference. Marking your working pipets can get messy, especially when the markings dissolve into your product. And scratching a working pipet with a file is asking for trouble. Get the scratch wet, a little lateral pressure and—snap—that's the end of the pipet.

Operation

I know I'm not supposed to have experiments in this book, but here's something I strongly suggest you carry out (Translation: Do It!). Work in a hood. Wear goggles. Don't point the end of a pipet at anyone—yourself included.

Experiment: Get a few millimeters of diethyl ether in a test tube. Fit a Pasteur pipet with a rubber bulb. Warm the pipet in your hand about one minute. Now draw some ether up into the pipet and watch.

If this went at all well, the ether probably came squirting out the end of the pipet. The ether evaporated in the pipet; then vapor filled the bulb and pushed the liquid out. Lots of liquids can do this, and if you've never pipetted anything but water before, the behavior of these liquids is quite striking.

Amelioration

There are two "fixes" to the vapor pressure problem:

1. Pre-wet the pipet with the solvent you'll have in there. Pull in and then expel the solvent two or three times. Once this vapor fills the pipet, the loss of solvent (as previously noted) is much slower, if not stopped. Even so, **be careful**. Don't squeeze or release the bulb rapidly; don't let your hand pre-heat or heat the pipet while handling it.
2. Stick a cork in it. A cotton plug really. Mayo et al. suggest that you push a small cotton ball way down into the narrow part of the pipet with a copper wire, rinse the cotton with 1 ml of methanol and then 1 ml of hexane, and let the cotton dry before use (Fig. 41a). On **all** your pipets. Well, sure, once you get good at it, it won't take all lab period to prepare them. Of course, you could just get good at being careful with this solvent pressure phenomenon.

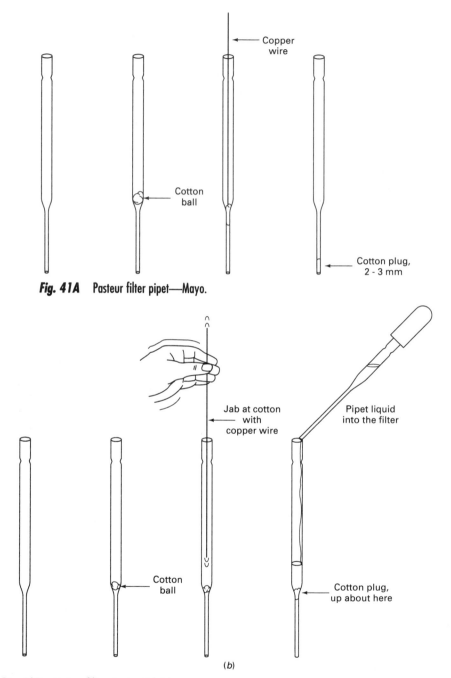

Fig. 41A Pasteur filter pipet—Mayo.

Fig. 41B Pasteur filter pipet—Zubrick.

I'm told a second advantage of putting cotton in all the pipets is that every time a transfer is made, the material is automatically filtered. Pull material into the pipet, and filth gets trapped on the cotton plug. Squirt the material out, and **all** the filth stays on the cotton. None **ever** washes back out. Yep. Sure. I believe, I believe....In any case, don't use too much cotton! A little here goes a long way.

PIPET CUTTING

From time to time you might need to shorten the long tip of the Pasteur pipet so that the narrow portion is, say, 2–3 cm rather than the 8 cm or so from the 9 inch pipet.

1. Get a sharp triangular file or, better yet, a sharp glass scorer.
2. Support the narrow part of the tip on a raised solid surface such as a ringstand plate.
3. Carefully draw the cutter back toward you while you rotate the pipet away from you. Don't press down too much; you'll crush the tip.
4. Wet the scratch.
5. Using a towel to hold the pipet and to keep from cutting yourself (although the small diameter of the tip makes disaster a bit less likely), position your thumbs to either side of the scratch, with the scratch facing away from you.
6. Read the next two steps before doing anything.
7. Exert force to pull the pipet apart. That's right. Pull first. (The pipet might part here.)
8. If it's still together, while you're **still pulling**, push both thumbs outward.
9. If it hasn't parted yet, either get help from your instructor or don't fool with this one anymore. Get another pipet and try again.
10. Carefully fire-polish the cut end.

PIPET FILTERING—LIQUIDS

Real pipet filtering will trap all of the solid materials. For this you need two pipets.

1. Stuff a cotton ball down into the pipet. Then use a wire to repeatedly tamp the cotton into the pipet. Give it 6 to 10 sharp jabs—much as if you were killing the nerves in a tooth during a root canal (Fig. 41b). You don't have to push the cotton all the way into the narrow part of the tip. Just wedge it firmly in the narrows.
2. Get a vial or other receptacle ready to receive your filtered product.
3. Fill the filter pipet from another pipet.
4. Let the liquid drain out of the filter pipet into your clean vial. Note that you could—could, mind you—put a rubber bulb onto the filter pipet and use the bulb to help push the solution out. Watch out:

    a. If you push the solution through the cotton too rapidly, you could force filth through the cotton and into the cleaned product.
    b. Don't *ever* let up on the rubber bulb once you start. You'll suck air and possibly filth off the cotton back into the pipet. In a few cases, where people didn't jam the cotton into the tip, but sort of had the cotton as a large ball wedged in the narrowing part, the cotton would be sucked back into the pipet, and the entire solution, filth and all, would run right out of the pipet.

5. You could—this is rare—blow the cotton jammed in the tip right out.

PIPET FILTERING—SOLIDS

After a recrystallization, you usually collect the new crystals by suction on a Buchner funnel (see Chapter 13, "Recrystallization"). For microscale quantities you may have to use a Hirsch funnel—a tiny Buchner funnel with sloping sides and a flat porous plate (see Chapter 5, "Other Interesting Equipment"). Or you might need the high-tech power of Craig tubes (see Chapter 14, "Recrystallization: Microscale"). Or you might be able to get away with a Pasteur pipet (Fig. 42).

1. Get the vial or tube containing the ice-cold crystalline mixture and stir it a bit with a Mayo-type cotton-plugged pipet—the cotton plug right to the end of the tip.
2. Slowly—there's that word again—push air out of the pipet while you bring the tip to the bottom of the container.
3. Slowly—there's that word yet again—draw solvent into the pipet, leaving crystals behind.

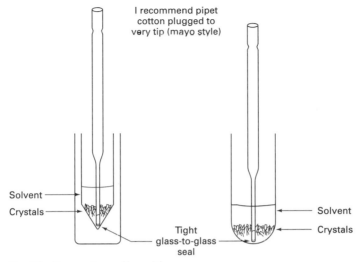

I recommend pipet
cotton plugged to
very tip (mayo style)

Solvent

Crystals

Tight
glass-to-glass
seal

Solvent

Crystals

Fig. 42 Pasteur pipet to filter solids.

Now you might want to pack the crystals down a bit by rapping the container on a hard surface. Then repeat the filtration above.

You can wash these crystals with a few drops of cold, fresh solvent and then remove the solvent as many times as is necessary. Eventually, however, you'll have to take the crystals out of the tube and let them dry.

If the tube with crystals has a flattened or even a slightly rounded bottom, you might be able to get away with using an unplugged pipet. The square end of the pipet can fit even a test tube bottom fairly closely, letting solvent in and keeping solvent out. With conical vials, there's more trouble getting a close glass-to-glass contact (not impossible, though), and it might be easier to stick to the Mayo-type plugged pipet (Fig. 42).

In all cases, make sure you keep the crystallization tube cold and that you wait, even the small length of time it takes, for the pipet tip to cool as well. You don't want the pipet tip warming the solution.

SYRINGES
AND NEEDLES

8

Another misgiving I have about microscale is that you often have to handle samples by syringe. There are about 25 rules for handling syringes. The first five are:

1. These are dangerous—watch it.
2. Watch it—don't stick yourself.
3. Don't stick anyone else.
4. Be careful.
5. Don't fool around with these things.

The other 20 rules—well, you get the idea. If that's not enough, some states require *extremely* tight control of syringes and needles. They might be kept under double lock and key in a locked cabinet in a locked room, and require extensive sign-out *and* sign-in procedures. And if you do get stuck, bring yourself and the syringe to the instructor immediately.

You'll probably use a glass or plastic syringe with a **Luer tip** (Fig. 43). This is a standard shape tip that can fit various pieces of apparatus, including the syringe needle. All you do is push the tip into the receptacle. Don't confuse this with the **Luer lock**, a special collar placed around the

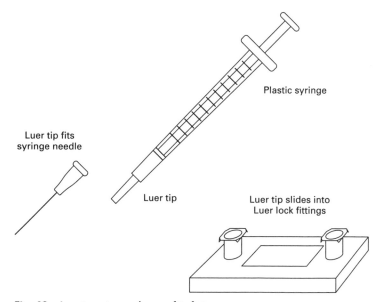

Plastic syringe

Luer tip fits
syringe needle

Luer tip

Luer tip slides into
Luer lock fittings

Fig. 43 Luer tip syringe and some of its fittings.

Luer tip designed to hold, well, *lock* the syringe barrel onto the apparatus, including needles (Fig. 44). You push the lock onto the receptacle and, with a twist, lock the syringe barrel to it. You'll have to *untwist* this beast to release the syringe barrel.

The biggest mistake—one that I still make—is losing my patience when filling a syringe with a needle attached. I always tend to pull the plunger quickly, and I wind up sucking air into the barrel. Load the syringe slowly. If you do get air in the barrel, push the liquid back out and try again.

Injecting a sample through a Teflon-lined plastic disk, or a rubber septum, is fairly straightforward. Don't get funny and try slinging the needle—medical school style—through the disk. Use two hands if you have to: one to hold the barrel and the other to keep the plunger from moving. Slowly push the plunger into the barrel to deliver the liquid.

THE RUBBER SEPTUM

If you're using syringes, you should know about the **rubber septum**. You put the smaller end into the opening you're going to inject into, and then roll the larger end over the *outside* of the opening (Fig. 45). You

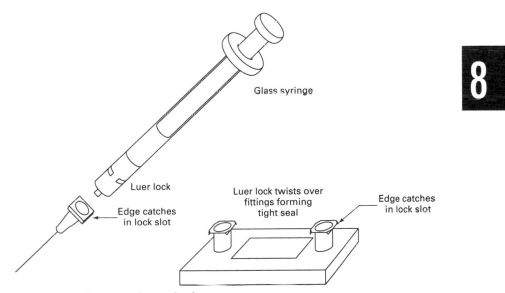

Fig. 44 Luer lock syringe and some of its fittings.

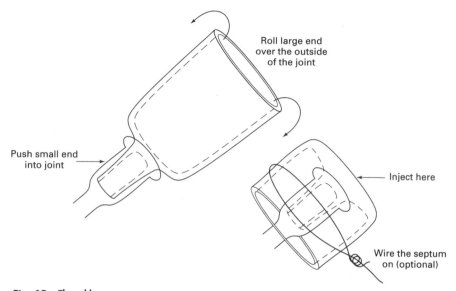

Roll large end
over the outside
of the joint

Push small end
into joint

Inject here

Wire the septum
on (optional)

Fig. 45 The rubber septum.

might use a length of copper wire to keep the rubber septum from coming out of the opening, especially if there's any chance of a pressure buildup inside the apparatus.

CLEAN AND DRY

Once you've identified your apparatus, you may find you have to clean it.

1. Wash your glassware at the *end* of the lab day. That way you'll have clean and dry glassware, ready to go for the next lab. This may be difficult to do if you perform an experiment on the day you check in.

2. A *little* solvent, a *little* detergent, and a *lot* of elbow grease. These are the correct proportions for a cleaning solution. You do not need all the soap on the planet, nor do you have to fill the glassware to overflowing with soap solution. *Agitation* is the key here. The more you agitate a small amount of soap solution, the less you agitate your instructor by wasting your time and supplies, and the more effective your cleaning will be.

3. Special Buchner funnel cleaning alert. The standard ceramic Buchner funnel is not transparent, and you can't see whether or not the bums who used the funnel the last time to collect a highly colored product didn't clean the funnel properly. The first time you Buchner filter crystals from an alcohol solution, the colored impurity dissolves, bleeds up into your previously clean crystals and you may have to redo your entire experiment. I'd rinse the Buchner funnel with a bit of hot ethanol before I used it, just for insurance.

DRYING YOUR GLASSWARE WHEN YOU DON'T NEED TO

"It's late. Why haven't you started the experiment yet?"
"I washed all my glassware and spent half an hour drying it."
"What technique are we doing?"
"Steam distillation."
"Steam goes through the entire setup, does it?"
A nodding head responds.
"What's condensed steam?"
"Water...."

There are all sorts of variations, but they boil down to this: You've taken all this time to dry your glassware only to put water in it. Writers of lab manuals are very tricky about this. Perhaps they say you'll be using *steam*. Or maybe 5% *aqueous* sodium bicarbonate solution. Or even that a byproduct of your reaction is H_2O. Condense *steam* and you get *what*? An *aqueous* solution has *what* for a solvent? H_2O is *what*?

Look for sources of water other than plain water. If a "water-and" mixture is going to be in the equipment anyway, drying to perfection is silly.

DRYING YOUR GLASSWARE WHEN YOU NEED TO

If you wash your glassware *before* you quit for the day, the next time you need it, it'll be clean and dry. There are only a few reactions you might do that need superclean, superdry apparatus, and you should be given special instructions when that's necessary. (In their book, *Experimental Organic Chemistry*, 2nd edition, McGraw-Hill, 1986, authors H. D. Durst and G. W. Gokel claim that glassware dried overnight is dry enough for the Grignard reaction, an *extremely* moisture-sensitive reaction, and that flame drying can be avoided unless the laboratory atmosphere is extremely humid.)

Don't use the compressed air from the compressed air lines in the lab for drying anything. These systems are full of dirt, oil, and moisture from the pumps, and they will get your equipment dirtier than before you washed it.

Yes, there are a few quick ways of drying glassware in case of emergency. You can rinse *very wet* glassware with a small amount of acetone, *drain the glassware very well*, and put the glassware in a drying oven (about 100°C) for a short spell. The acetone not only washes the water off the glassware very well (the two liquids are **miscible**, that is, they mix in all proportions), but also the liquid left behind is acetone-rich and evaporates faster than water. Don't use this technique unless absolutely necessary.

DRYING AGENTS

When you've prepared a liquid product, you must dry the liquid before you finally distill and package it, by treating the liquid with a **drying agent**. Drying agents are usually certain anhydrous salts that combine with the water in your product and hold it as a **water of crystallization**. When all the water in your sample is tied up with the salt, you gravity filter the mixture. The dried liquid passes through the filter paper and the *hydrated salt* stays behind.

TYPICAL DRYING AGENTS

1. *Anhydrous calcium chloride.* This is a very popular drying agent, inexpensive and rapid, but of late I've become disappointed in its performance. It seems that the calcium chloride powders a bit upon storage and abuse, and this *calcium chloride dust* can go right through the filter paper with the liquid. So a caution: If you must use anhydrous calcium chloride, be sure it is granular. Avoid powdered calcium chloride, or granular anhydrous calcium chloride that's been around long enough to become pulverized. And don't add to the problem by leaving the lid off the jar of the drying agent; that's the abuse I was talking about.

   Anhydrous calcium chloride tends to form *alcohols of crystallization,* so you really can't use it to dry alcohols.

2. *Anhydrous sodium carbonate and anhydrous potassium carbonate.* These are useful drying agents that are basically *basic.* As they dry your organic compound, any carbonate that gets dissolved in the tiny amounts of water in your sample can neutralize any tiny amounts of acid that may be left in the liquid. If your product is *supposed* to be acidic (in contrast to being *contaminated* with acid), you should avoid these drying agents.

3. *Anhydrous magnesium sulfate.* In my opinion, anh. $MgSO_4$ is about the best all-around drying agent. It has a drawback, though. Since it is a fine powder, lots of your water can become trapped on the surface. *This is not the same as water of crystallization.* The product is *only on the surface, not inside the crystal structure,* and you may wash your product off.

4. *Drierite.* Drierite, a commercially available brand of anhydrous calcium sulfate, has been around a long time and is a popular drying agent. You can put it in liquids and dry them or pack a drying tube

with it to keep the moisture in the air from getting into the reaction setup. But be warned. There is also *blue Drierite*. This has an indicator, a cobalt salt, that is *blue when dry, pink when wet*. Now you can easily tell when the drying agent is no good. Just look at it. Unfortunately, this stuff is not cheap, so don't fill your entire drying tube with it just because it'll look pretty. Use a small amount mixed with white Drierite, and when the blue pieces turn pink, change the entire charge in your drying tube. You can take a chance using blue Drierite to dry a liquid directly. Sometimes the cobalt compound dissolves in your product. Then you have to clean and dry your product all over again.

USING A DRYING AGENT

1. Put the liquid or solution to be dried into an Erlenmeyer flask.
2. Add *small* amounts of drying agent and swirl the liquid. When the liquid is no longer cloudy, the water is gone, and the liquid is dry.
3. Add just a bit more drying agent and swirl one final time.
4. Gravity filter through filter paper (see Fig. 59).
5. If you've used a carrier solvent, then evaporate or distill it off, whichever is appropriate. Then you'll have your clean, dry product.

FOLLOWING DIRECTIONS AND LOSING PRODUCT ANYWAY

"Add 5 g of anhydrous magnesium sulfate to dry the product." Suppose your yield of product is lower than that in the book. Too much drying agent—not enough product—Zap! It's all sucked onto the surface of the drying agent. Bye bye product. Bye bye grade.

Add the drying agent slowly to the product in small amounts.

Now about those small amounts of product (usually liquids).

1. Dissolve your product in a *low boiling point solvent*. Maybe ether or hexane or the like. Now dry this whole solution, and gravity filter. Remove the solvent carefully. Hoo-ha! Dried product.
2. Use chunky dehydrating agents like anhydrous calcium sulfate (Drierite). Chunky drying agents have a much smaller surface area, so not much of the product gets absorbed.

10

DRYING AGENTS: MICROSCALE

Using drying agents in microscale work follows the same rules as drying on larger scales. Chunky Drierite becomes a problem, and you may have to chop it up a bit (good luck!) before you use it. And, as ever, beware directions that command you to fill a cotton-plugged pipet (see Chapter 7, "Pipet Tips") with exactly 2.76 g of drying agent: again, this is a *typical* amount of drying agent to be used to dry the *typical* amount of product. The one nice thing about drying your liquids this way is that you can also filter impurities out of the liquid at the same time (Fig. 41b).

ON PRODUCTS

The fastest way to lose points is to hand in messy samples. Lots of things can happen to foul up your product. The following are unforgivable sins!

SOLID PRODUCTS

1. ***Trash in the sample.*** Redissolve the sample, gravity filter, and then evaporate the solvent.
2. ***Wet solids.*** Press out on filter paper, break up, and let dry. The solid shouldn't stick to the sides of the sample vial. Tacky!
3. ***Extremely wet solids (solid floating in water).*** Set up a gravity filtration (see Chapter 13, "Recrystallization") and filter the liquid off of the solid. Remove the filter paper cone with your solid product, open it up, and leave it to dry. Or remove the solid and dry it on fresh filter paper as above. Use lots of care, though. You don't want filter paper fibers trapped in your solid.

LIQUID PRODUCTS

1. ***Water in the sample.*** This shows up as droplets or as a layer of water on the top or the bottom of the vial, or *the sample is cloudy*. Dry the sample with a drying agent (see Chapter 10, "Drying Agents") and gravity filter into a clean dry vial.
2. ***Trash floating in the sample.*** For that matter, it could be on the bottom, lying there. Gravity filter into a clean, dry vial.
3. ***Water in the sample when you don't have a lot of sample.*** Since solid drying agents can absorb lots of liquid, what can you do if you have a tiny amount of product to be dried? Add some solvent that has a low boiling point. It must dissolve your product. Now you have a lot of liquid to dry, and *if a little gets lost, it is not all product*. Remove this solvent after you've dried the solution. Be careful if the solvent is flammable. *No flames!*

THE SAMPLE VIAL

Sad to say, but an attractive package can sell an inferior product. So why not sell yours. Dress it up in a **neat new label**. Put on

1. **Your name.** Just in case the sample gets lost on the way to camp.
2. **Product name.** So everyone will know what is in the vial. What does "Product from part C" mean to you? Nothing? Funny, it doesn't mean anything to instructors either.
3. **Melting point (solids only).** This is a range, like "M.P. 96–98°C" (see Chapter 12, "The Melting Point Experiment").
4. **Boiling point (liquids only).** This is a range "B.P. 96–98°C" (see Chapter 20, "Distillation").
5. **Yield.** If you weigh the empty vial and cap, you have the **tare**. Now add your product and weigh the full vial. Subtract the *tare* from this *gross weight* to get the **net weight** (yield, in grams) of your product.
6. **Percent yield.** Calculate the percent yield (see Chapter 2, "Keeping a Notebook") and put it on the label.

You may be asked for more data, but the things listed above are a good start down the road to good technique.

P.S. Gummed labels can fall off vials, and pencil will smear. *Always use waterproof ink!* And a piece of transparent tape over the label will keep it on.

HOLD IT! DON'T TOUCH THAT VIAL

Welcome to "You Bet Your Grade." The secret word is **dissolve**. Say it slowly as you watch the cap liner in some vials dissolve into your nice, clean product and turn it all goopy. This can happen. A good way to prevent it is to cover the vial with aluminum foil before you put the cap on. Just make sure the product does not react with aluminum. Discuss this procedure at length with your instructor.

THE MELTING POINT EXPERIMENT

A **melting point** is the temperature at which the first crystal just starts to melt until the temperature at which the last crystal just disappears. Thus the melting point (abbreviated M.P.) is actually a **melting range**. You should report it as such, even though it is *called* a melting point, for example, M.P. 147–149°C.

People always read the phrase as melting *point* and never as *melting* point. There is this uncontrollable, driving urge to report one number. No matter how much I've screamed and shouted at people *not* to report one number, they almost always do. It's probably because handbooks list only one number, the upper limit.

Generally, melting points are taken for two reasons.

1. ***Determination of purity.*** If you take a melting point of your compound and it starts melting at 60°C and doesn't finish until 180°C, you might suspect something is wrong. A melting range *greater than* 2°C usually indicates an impure compound. (As with all rules, there are exceptions. There aren't many to this one, though.)

2. ***Identification of unknowns.***

   a. If you have an unknown solid, take a melting point. Many books (ask your instructor) contain tables of melting points and lists of compounds that may have a particular melting point. One of them may be your unknown. You may have 123 compounds to choose from. A little difficult, but that's not all the compounds in the world. Who knows?? Give it a try. If nothing else, you know the melting point.

   b. Take your unknown and mix it *thoroughly* with some chemical you think might be your unknown. You might not get a sample of it, but you can ask. Shows you know something. Then:

   (1) If the mixture melts at a *lower* temperature, over a *broad range*, your unknown is not the same compound.

   (2) If the mixture melts at the *same temperature, same range*, it's a good bet it's the *same compound*. Try another one, though, with a different ratio of your unknown and this compound just to be sure. A *lower* melting point with a *sharp range* would be a special point called a **eutectic mixture**, and you, with all the other troubles in lab, just might accidentally hit it. On lab quizzes, this is called

   "Taking a mixed melting point."

Actually, "taking a mixture melting point," the melting point of a mixture, is more correct. But I have seen this expressed both ways.

SAMPLE PREPARATION

You usually take melting points in thin, closed end tubes called **capillary tubes**. They are also called **melting point tubes** or even **melting point capillaries**. The terms are interchangeable, and I'll use all three.

Sometimes you may get a supply of tubes that are open on *both ends*! You don't just use these as is. Light a burner and before you start, close off one end. Otherwise your sample will fall out of the tube (see "Closing Off Melting Point Tubes" later in this chapter).

Take melting points on *dry, solid* substances only, *never* on liquids or solutions of solids *in* liquids or on wet or even damp solids. Only on dry solids!

To help dry damp solids, place the damp solid on a piece of filter paper and fold the paper around the solid. Press. Repeat until the paper doesn't get wet. Yes, you may have to use fresh pieces of paper. Try not to get filter paper fibers in the sample, OK?

Occasionally, you may be tempted to dry solid samples in an oven. *Don't*—unless you are specifically instructed to. I know some students who have decomposed their products in ovens and under heat lamps. With the time they save quickly decomposing their product, they can repeat the entire experiment.

Loading the Melting Point Tube

Place a small amount of *dry* solid on a new filter paper (Fig. 46). Thrust the open end of the capillary tube into the middle of the pile of material. Some solid should be trapped in the tube. Turn the tube over, closed end down. Remove any solid sticking to the outside. The solid must now be packed down.

Traditionally, the capillary tube, turned upright with the *open end up*, is stroked with a file, or tapped on the benchtop. Unless done *carefully*, these operations *may break the tube*. A safer method is to drop the tube *closed end down*, through a length of glass tubing. You can even use your condenser or distilling column for this purpose. When the capillary strikes the benchtop, the compound will be forced into the closed end. You may have to do this several times. If there is not enough material in the M.P. tube,

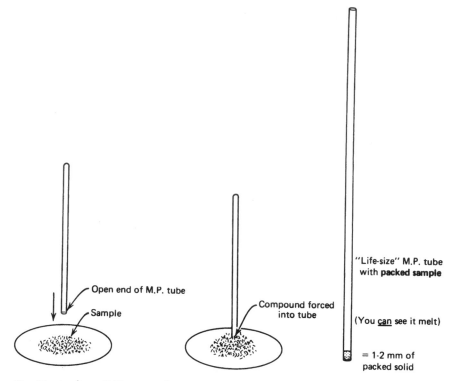

Open end of M.P. tube

Sample

Compound forced
into tube

"Life-size" M.P. tube
with **packed sample**

(You <u>can</u> see it melt)

= 1-2 mm of
packed solid

Fig. 46 Loading a Melting point tube.

thrust the open end of the tube into the mound of material and pack it down
again. Use your own judgment; consult your instructor.

Use the smallest amount of material that can be seen to melt.

Closing Off Melting Point Tubes

If you have melting point tubes that are *open at both ends* and you try
to take a melting point with one, it should come as no surprise when your
compound falls out of the tube. You'll have to *close off one end* to keep your
sample from falling out (Fig. 47). So light a burner and get a "stiff" small
blue flame. Slowly touch the end of the tube to the side of the flame, and hold
it there. You should get a yellow sodium flame, and the tube will close up.
There is no need to rotate the tube. And remember, *touch—just touch—*the

edge of the flame, and hold the tube there. Don't feel you have to push the tube way into the flame.

12

MELTING POINT HINTS

1. Use only the smallest amount that you can see melt. Larger samples will heat unevenly.
2. Pack down the material as much as you can. Left loose, the stuff will heat unevenly.
3. Never remelt any sample. They may undergo nasty chemical changes such as oxidation, rearrangement, and decomposition.
4. Make up more than one sample. One is easy, two is easier. If something goes wrong with one, you have another. Duplicate, even triplicate runs are common.

THE MEL-TEMP APPARATUS

The Mel-Temp apparatus (Fig. 48) substitutes for the Thiele tube or open beaker and hot oil methods (see "Using the Thiele Tube" later in this chapter). Before you use the apparatus, there are a few things you should look for.

1. *Line cord.* Brings AC power to the unit. Should be plugged into a live wall socket. [See J. E. Leonard and L. E. Mohrmann, *J. Chem. Educ.*, **57**, 119 (1980), for a modification in the wiring of older units, to make them less lethal. It seems that even with the three-prong plug, there can still be a shock hazard. *Make sure your instructor knows about this!*]
2. *On-off switch.* Turns the unit on or off.

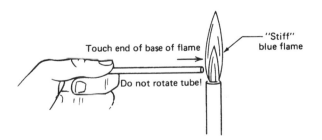

Fig. 47 Closing off a M.P. tube with a flame.

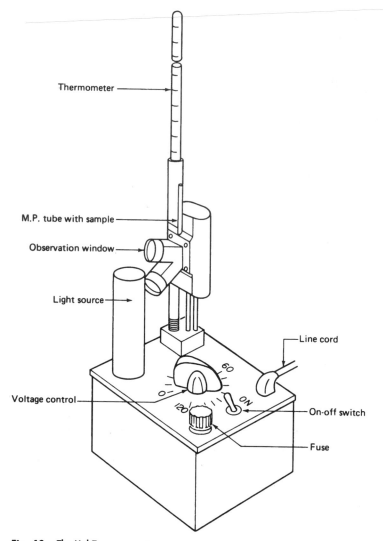

Fig. 48 The Mel-Temp apparatus.

3. **Fuse.** Provides electrical protection for the unit.
4. **Voltage control.** Controls the *rate* of heating, *not the temperature!* The higher the setting, the faster the temperature rise.
5. **Light source.** Provides illumination for samples.
6. **Eyepiece.** Magnifies the sample (Fig. 49).
7. **Thermometer.** Gives the temperature of the sample, and upsets the digestion when you're not careful and you snap it off in the holder.

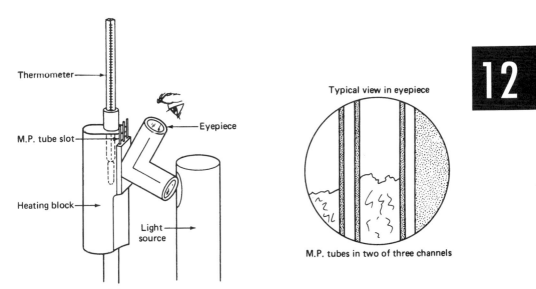

Fig. 49 Close-up of the viewing system.

OPERATION OF THE MEL-TEMP APPARATUS

1. *Imagine yourself getting burned if you're not careful.* Never assume the unit is cold.
2. Place loaded M.P. tube in one of the three channels in the opening at the top of the unit (Fig. 49).
3. Set the voltage control to zero if necessary. There are discourteous folk who do not reset the control when they finish using the equipment.
4. Turn the on-off switch to ON. The light source should illuminate the sample. If not, call for help.
5. Now science turns into art. Set the *voltage control* to *any* convenient setting. The point is to get up to *within 20°C* of the *supposed* melting point. Yep, that's right. If you have no idea what the melting point is, it may require several runs as you keep skipping past the points with a temperature rise of 5-10°C per minute. A convenient setting is 40. This is just a suggestion, not an article of faith.
6. After you've melted a sample, *throw it away!*
7. Once you have an idea of the melting point (or looked it up in a handbook, or were told), *get a fresh sample*, and bring the temperature up quickly at about *5-10°C per minute C per minute* to *within 20°C* of this approximate melting point. Then turn down the *voltage control* to get a 2°C *per minute rise*. Patience!

8. When the first crystals *just start to melt*, record the temperature. When the *last crystal just disappears*, record the temperature. If both points appear to be the same, either the sample is extremely pure or the temperature rise was *too fast*.
9. Turn the on-off switch to OFF. You can set the voltage control to zero for the next person.
10. Remove all capillary tubes.

Never use a wet rag or sponge to quickly cool off the heating block. This might permanently warp the block. You can use a cold metal block to cool it if you're in a hurry. Careful. If you slip, you may burn yourself.

THE FISHER-JOHNS APPARATUS

The Fisher-Johns apparatus (Fig. 50) is different in that you don't use capillary tubes to hold the sample. Instead, you sandwich your sample between two round microscope cover slides (thin windows of glass) on a heating block. This type of melting point apparatus is called a **hot stage**. It comes complete with spotlight. Look for the following.

1. *Line cord* (at the back). Brings AC power to the unit. Should be plugged into a live wall socket.
2. *On-off switch.* Turns the unit on or off.
3. *Fuse* (also at the back). Provides electrical protection for the unit.
4. *Voltage control.* Controls the *rate* of heating, *not the temperature!* The higher the setting, the faster the temperature rise.
5. *Stage light.* Provides illumination for samples.

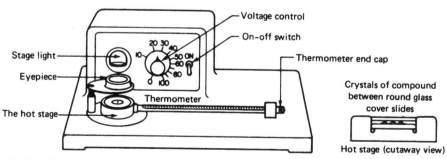

Fig. 50 The Fisher-Johns Apparatus.

6. **Eyepiece.** Magnifies the sample.
7. **Thermometer.** Gives the temperature of sample.
8. **Thermometer end cap.** Keeps the thermometer from falling out. If the cap becomes loose, the thermometer tends to go belly-up and the markings turn over. Don't try to fix this while the unit is hot. Let it cool so you won't get burned.
9. **The hot stage.** This is the heating block that samples are melted on.

OPERATION OF THE FISHER-JOHNS APPARATUS

1. *Don't assume that the unit is cold.* That is a good way to get burned.
2. Keep your grubby fingers off the cover slides. Use tweezers or forceps.
3. Place a clean round glass cover slide in the well on the hot stage. *Never melt any samples directly on the metal stage.* Ever!
4. Put a few crystals on the glass. Not too many. As long as you can see them melt, you're all right.
5. Put another cover slide on top of the crystals to make a sandwich.
6. Set the voltage control to zero if it's not already there.
7. Turn on-off switch to ON. The light source should illuminate the sample. If not, call for help!
8. Now science turns into art. Set the *voltage control to any* convenient setting. The point is to get up to *within 20°C* of the *supposed* melting point. Yep, that's right. If you have no idea what the melting point is, it may require several runs as you keep skipping past the point with a temperature rise of 5-10°C per minute. A convenient setting is *40*. This is just a suggestion, not an article of faith.
9. After you've melted a sample, let it cool, and remove the sandwich of sample and cover slides. *Throw it away!* Use an appropriate waste container.
10. Once you have an idea of the melting point (or have looked it up in a handbook, or you were told), *get a fresh sample*, and bring the temperature up quickly at about *5-10°C per minute to within 20°C* of this approximate melting point. Then turn down the *voltage control* to get a *2°C per minute rise*. Patience!
11. When the first crystals *just start to melt*, record the temperature. When the last crystal *just disappears*, record the temperature. If both points appear to be the same, either the sample is extremely pure or the temperature rise was *too fast*.

12. Turn the on-off switch to OFF. Now set the voltage control to zero.
13. Let the stage cool, then remove the sandwich.

THE THOMAS-HOOVER APPARATUS

The Thomas-Hoover apparatus (Fig. 51) is the electromechanical equivalent of the Thiele tube or open beaker and hot oil methods (see "Using the Thiele Tube" later in this chapter). It has lots of features, and you should look for the following.

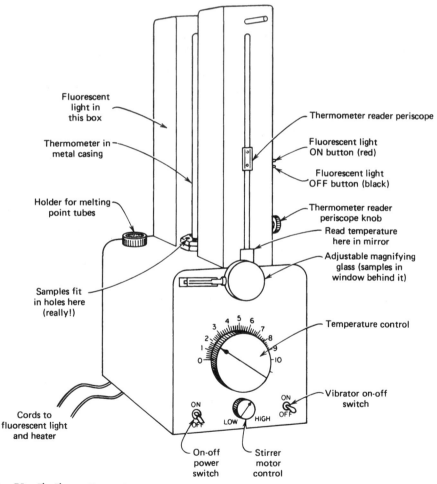

Fluorescent light in this box

Thermometer in metal casing

Holder for melting point tubes

Samples fit in holes here (really!)

Cords to fluorescent light and heater

Thermometer reader periscope

Fluorescent light ON button (red)

Fluorescent light OFF button (black)

Thermometer reader periscope knob

Read temperature here in mirror

Adjustable magnifying glass (samples in window behind it)

Temperature control

Vibrator on-off switch

On-off power switch

Stirrer motor control

Fig. 51 The Thomas-Hoover Apparatus.

12

1. *Light box.* At the top of the device, toward the back, a box holds a fluorescent light bulb behind the thermometer. On the right side of this box are the fluorescent light switches.
2. *Fluorescent light switches.* Two buttons. Press and hold the red button down for a bit to light the lamp; press the black button to turn the lamp off.
3. *Thermometer.* A special 300° thermometer in a metal jacket is immersed in the oil bath that's in the lower part of the apparatus. Two slots have been cut in the jacket to let light illuminate the thermometer scale from behind and to let a thermometer periscope read the thermometer scale from the front.
4. *Thermometer periscope.* In front of the thermometer, this periscope lets you read a small magnified section of the thermometer scale. By turning the small knob at the lower right of this assembly, you track the movement of the mercury thread, and an image of the thread and temperature scale appears in a stationary mirror just above the sample viewing area.
5. *Sample viewing area.* A circular opening is cut in the front of the metal case so that you can see your samples in their capillary tubes (and the thermometer bulb) all bathed in the oil bath. You put the tubes into the oil bath through the holes in the capillary tube stage.
6. *Capillary tube stage.* In a semicircle about the bottom of the jacketed thermometer, yet behind the thermometer periscope, are five holes through which you can put your melting point capillaries.
7. *Heat.* Controls the rate of heating, not the temperature. The higher the setting, the faster the temperature rise. At Hudson Valley Community College, we've had a stop put in and you can only turn the dial as far as the number 7. When it gets up to 10, you always smoke the oil. Don't do that.
8. *Power on-off switch.* Turns the unit on or off.
9. *Stirrer control.* Sets the speed of the stirrer from low to high.
10. *Vibrator on-off switch.* Turns the vibrator on or off. It's a spring-return switch, so you must hold the switch in the ON position. Let go, and it snaps off.
11. *Line cords.* One cord brings AC power to the heater, stirrer, sample light, and vibrator. The other cord brings power to the fluorescent light behind the thermometer. Be sure both cords are plugged into live wall sockets.

OPERATION OF THE THOMAS-HOOVER APPARATUS

1. If the fluorescent light for the thermometer is not lit, press the red button at the right side of the light box and hold it down for a bit to start the lamp. The lamp should remain lit after you release the button.

2. Look in the thermometer periscope, turn the small knob at the lower right of the periscope base, and adjust the periscope to find the top of the mercury thread in the thermometer. Read the temperature. Wait for the oil bath to cool if the temperature is fewer than 20 Celsius degrees below the approximate melting point of your compound. You'll have to wait for a room temperature reading if you have no idea what the melting point is. You don't want to plunge your sample into oil that is so hot it might melt too quickly, or at an incorrect temperature.

3. Turn the voltage control to zero if it isn't there already.

4. Turn the power on-off switch to ON. The oil bath should become illuminated.

5. Insert your capillary tube in one of the capillary tube openings in the capillary tube stage. *This is not simple.* Be careful. If you snap a tube at this point, the entire unit may have to be taken apart to remove the pieces. It appears you have to angle the tube toward the center opening and angle the tube toward you (as you face the instrument) at the same time (Fig. 52). It's as if they were placed on the surface of a conical funnel.

6. Adjust the magnifying glass for the best view of your sample.

7. Turn the stirrer knob so that the mark on the knob is about halfway between the slow and fast markings on the front panel. That's just a suggestion. I don't have any compelling reasons for it.

8. Adjust the thermometer periscope to give you a good view of the top of the mercury thread in the thermometer.

9. Now science turns into art. Set the *heat control* to *any* convenient setting. The point is to get up to *within 20°C* of the *supposed* melting point. If you have no idea what the melting point is, it may require several runs as you keep skipping past the point with a temperature rise of 5–10°C per minute. A convenient setting is *4*. This is just a suggestion, not an article of faith.

10. Remember, you'll have to keep adjusting the thermometer periscope to keep the top of the mercury thread centered in the image.

11. After you've melted a sample, *throw it away!*

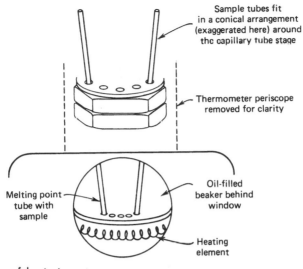

Sample tubes fit
in a conical arrangement
(exaggerated here) around
the capillary tube stage

Thermometer periscope
removed for clarity

Melting point
tube with
sample

Oil-filled
beaker behind
window

Heating
element

Fig. 52 Close-up of the viewing system.

12. Once you have an idea of the melting point (or looked it up in a handbook, or were told), *get a fresh sample* and bring the temperature up quickly at about *5–10°C* per minute to *within 20°C* of this approximate melting point. Then turn down the *heat control* to get a *2°C* per minute rise. Patience!

13. When the first crystals *just start to melt*, record the temperature. When the *last crystal just disappears*, record the temperature. If both points appear to be the same, either the sample is extremely pure or the temperature rise was *too fast*. If you record the temperature with the horizontal index line in the mirror matched to the lines etched on both sides of the periscope window and the top of the mercury thread at the same time, you'll be looking at the thermometer scale head on. This will give you the smallest error in *reading* the temperature (Fig. 53).

14. Don't turn the control much past 7. You can get a bit beyond 250°C at that setting, and that should be *plenty* for any solid compound you might prepare in this lab. Above this setting, there's a real danger of smoking the oil.

15. Turn the power switch to OFF. You can also set the *heat control* to zero for the next person.

16. Press the black button on the right side of the light box and turn the fluorescent light off.

17. Remove all capillary tubes.

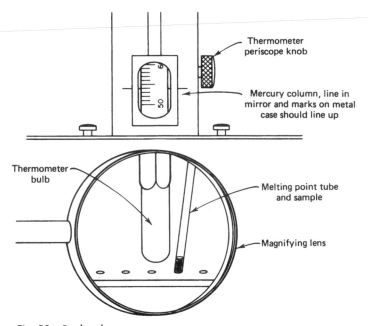

Fig. 53 Reading the temperature.

There are a few more electric melting point apparatuses around, and much of them work the same. A **sample holder**, **magnifying eyepiece**, and **voltage control** are common; an apparently essential feature of these devices is that dial markings are almost *never* temperature settings. That is, a setting of **60** will not give a temperature of 60°C but probably much higher.

USING THE THIELE TUBE

With the Thiele tube (Fig. 54), you use hot oil to transfer heat evenly to your sample in a melting point capillary, just like the metal block of the Mel-Temp apparatus does. You heat the oil in the sidearm and it expands. The hot oil goes up the sidearm, warming your sample and thermometer as it touches them. Now, the oil is cooler, and it falls to the bottom of the tube where it is heated again by a burner. This cycle goes on automatically as you do the melting point experiment in the Thiele tube.

Don't get any water in the tube or when you heat the tube the water can boil and throw hot oil out at you. Let's start from the beginning.

12

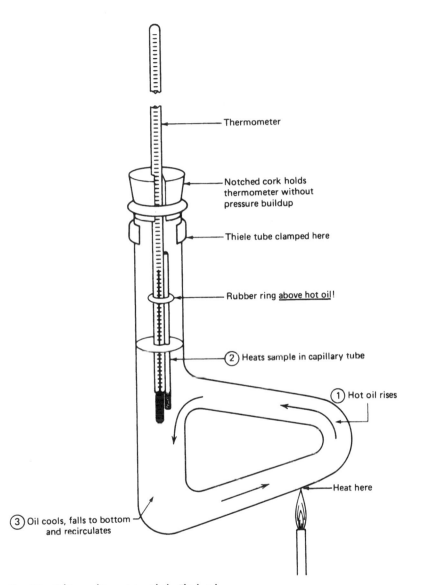

Thermometer

Notched cork holds
thermometer without
pressure buildup

Thiele tube clamped here

Rubber ring <u>above hot oil</u>!

(2) Heats sample in capillary tube

(1) Hot oil rises

Heat here

(3) Oil cools, falls to bottom
and recirculates

Fig. 54 Taking melting points with the Thiele tube.

Cleaning the Tube

This is a bit tricky, so don't do it unless your instructor says so. Also, check with your instructor *before* you put fresh oil in the tube.

1. Pour the old oil out into an appropriate container and let the tube drain.
2. Use a hydrocarbon solvent (hexane, ligroin, petroleum ether—and *no flames!*) to dissolve the oil that's left.
3. Get out the old soap and water and elbow grease, clean the tube, and rinse it out really well.
4. Dry the tube in a drying oven (usually >100°C) thoroughly. Carefully take it out of the oven and let it cool.
5. Let your instructor examine the tube. If you get the OK, *then* add some fresh oil. Watch it. First, *no water*. Second, don't overfill the tube. Normally, the oil expands as you heat the tube. If you've overfilled the tube, oil will crawl out and get you.

Getting the Sample Ready

Here you use a loaded melting point capillary tube (see "Loading the Melting Point Tube" earlier in the chapter) and attach it directly to the thermometer. Unfortunately, the thermometer has bulges; there are some problems, and you may snap the tube while attaching it to the thermometer.

1. Get, or cut, a thin rubber ring from a piece of rubber tubing.
2. Put the *bottom* of the loaded MP tube *just above* the place where the thermometer constricts (Fig. 55), and carefully roll the rubber ring onto the MP tube.
3. Reposition the tube so that the sample is near the center of the bulb and the rubber ring is near the open end. *Make sure the tube is vertical.*

Dunking the Melting Point Tube

There are more ways of keeping the thermometer suspended in the oil than I care to list. You can cut or file a notch on the side of the cork, drill a hole, and insert the thermometer *(be careful!)*. Finally, cap the Thiele tube (Fig.

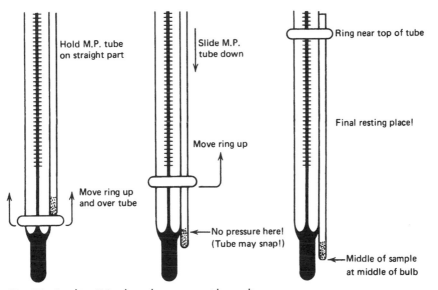

Fig. 55 Attaching M.P. tube to thermometer without a disaster.

54). The notch is there so that pressure will not build up as the tube is heated. *Keep the notch open, or the setup may explode.*

But this requires drilling or boring corks, something you try to avoid. (Why have ground-glass jointware in the undergraduate lab?) You can *gently* hold a thermometer and a cork in a clamp (Fig. 56). Not too much pressure, though!

Finally, you might put the thermometer in the **thermometer adapter** and suspend that, clamped gently by the rubber part of the adapter, and not by the ground glass end. Clamping ground glass will score the joint.

Heating the Sample

The appropriately clamped thermometer is set up in the Thiele tube as in Fig. 54. Look at this figure *now* and remember to heat the tube *carefully— always carefully—at the elbow. Then:*

1. If you don't know the melting point of the sample, heat the oil fairly quickly, *but no more than 10°C per minute* to get a rough melting point. And it will be rough indeed, since the temperature of the thermometer usually lags that of the sample.

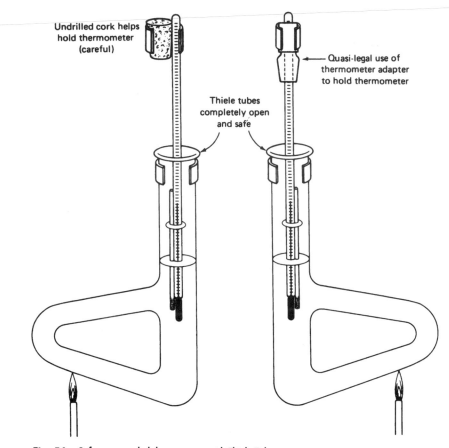

Undrilled cork helps
hold thermometer
(careful)

Quasi-legal use of
thermometer adapter
to hold thermometer

Thiele tubes
completely open
and safe

Fig. 56 Safety suspended thermometer with Thiele Tube.

2. After this sample has melted, lift the thermometer and attached sample tube *carefully (it may be HOT)* by the thermometer up at the clamp, until they are *just out of the oil*. This way the thermometer and sample can cool, and the hot oil can drain off. Wait for the thermometer to cool to about room temperature before you remove it entirely from the tube. Wipe off some of the oil, reload a melting point tube *(never remelt melted samples)*, and try again. And heat at 2°C per minute this time.

RECRYSTALLIZATION

The essence of a recrystallization is a **purification**. Messy, dirty compounds are cleaned up, purified, and can then hold their heads up in public again. The sequence of events you use will depend a lot on how messy your crude product is and on just how soluble it will be in various solvents.

In any case, you'll have to remember a few things.

1. Find a solvent that will *dissolve the solid while hot.*
2. The same solvent *should not dissolve it while cold.*
3. The *cold solvent* must keep impurities dissolved in it *forever or longer.*

This is the major problem. And it requires some experimentation. That's right! Once again, art over science. Usually, you'll know what you should have prepared, so the task is easier. It requires a trip to **your notebook** and, possibly, a **handbook** (see Chapter 2, "Keeping a Notebook," and Chapter 3, "Interpreting a Handbook"). You have the data on the solubility of the compound in your notebook. What's that you say? *You don't have the data in your notebook?* Congratulations, you get the highest F in the course.

Information in the notebook (which came from a handbook) for your compound might say, for alcohol (meaning *ethyl* alcohol), **s.h.** Since this means soluble in **h**ot alcohol, it implies **i**nsoluble in cold alcohol (and you wondered what the **i** meant). Then alcohol is probably a good solvent for recrystallization of that compound. Also, check on the **color** or **crystalline form**. This is important since

1. A color in a supposedly white product is an impurity.
2. A color in a colored product is *not* an impurity.
3. The *wrong color* in a product is an impurity.

You can usually assume impurities are present in small amounts. Then you don't have to guess what possible impurities might be present or what they might be soluble or insoluble in. If your sample is really dirty, the assumption can be fatal. This doesn't usually happen in an undergraduate lab, but you should be aware of it.

FINDING A GOOD SOLVENT

If the solubility data for your compound are not in handbooks, then

1. Place 0.1 g of your solid (weighed to 0.01 g) in a test tube.
2. Add 3 ml of a solvent, stopper the tube, and shake the bejesus out of it. If *all of the solid dissolves at room temperature*, then your solid is **soluble**. Do *not* use this solvent as a recrystallization solvent. (You must make note of this in your notebook, though.)
3. If none (or very little) of the solid dissolved at room temperature, unstopper the tube and heat it *(Careful— no flames and get a boiling stone!)* and shake it and heat it and shake it. You may have to heat the solvent to a gentle boil. (*Careful!* Solvents with low boiling points often boil away.) If it does *not* dissolve at all, then do not use this as a recrystallization solvent.

4. If the sample *dissolved when hot*, and *did not dissolve at room temperature*, you're on the trail of a good recrystallization solvent. One last test.
5. Place this last test tube in an ice-water bath, and cool it to about 5°C or so. If lots of crystals come out, this is good, and this is your recrystallization solvent.
6. Suppose your crystals don't come back when you cool the solution. Get a glass rod into the test tube, stir the solution, rub the inside of the tube with the glass rod, and agitate that solution. If crystals still don't come back, perhaps you'd better find another solvent.
7. Suppose, after all this, you still haven't found a solvent. Look again. Perhaps your compound *completely* dissolved in ethanol at room temperature and would *not* dissolve in water. AHA! Ethanol and water are **miscible** (i.e., they mix in all proportions) as well. You will have to perform a **mixed-solvent recrystallization** (see "Working with a Mixed-Solvent System" later in this chapter).

GENERAL GUIDELINES FOR A RECRYSTALLIZATION

Here are some general rules to follow for purifying any solid compound.

1. Put the solid in an Erlenmeyer flask, not a beaker. If you recrystallize compounds in beakers, you may find the solid climbing the walls of the beaker to get at you as a reminder. A 125 ml Erlenmeyer usually works. Your solid should look comfortable in it, neither cramped nor with too much space. You probably shouldn't fill the flask more than one fifth to one fourth full.

2. Heat a large quantity of a proven solvent (see preceding) to the boiling point, and *slowly add the hot solvent. Slowly!* A word about solvents: *Fire!* Solvents burn! *No flames!* A hot plate here would be better. You can even heat solvents in a *steam or water bath.* But—*no flames!*

3. Carefully add the hot solvent to the solid to just dissolve it. This can be tricky, since hot solvents evaporate, cool down, and so on. Ask your instructor.

4. Add a slight excess of the hot solvent (5–10 ml) to keep the solid dissolved.

5. If the solution is only slightly colored, the impurities will stay in solution. Otherwise, the big gun, **activated charcoal**, may be needed (see "Activated Charcoal" later in this chapter). Remember, if you were working with a colored compound, it would be silly to try to get rid of all the color, since you would get rid of all the compound and probably all your grade.

6. Keep the solvent hot *(not boiling)* and look carefully to see if there is any trash in the sample. This could be old boiling stones, sand, floor sweepings, and so on. Nothing you'd want to bring home to meet the folks. *Don't confuse real trash with undissolved good product!* If you add more hot solvent, good product will dissolve and trash will not. If you have trash in the sample, do a **gravity filtration** (see the following).

7. Let the Erlenmeyer flask and the hot solution cool. Slow cooling gives better crystals. Garbage doesn't get trapped in them. But this can take what seems to be an interminable length of time. (I know, the entire lab seems to take an interminable length of time.) So, after the flask cools and it's just *warm* to the touch, then put the flask in an ice-water bath to cool. *Watch it!* The flasks have a habit of turning over in the water baths and letting all sorts of water destroy all your hard work! Also, a really hot flask will shatter if plunged into the ice bath, so again, watch it.

8. When you're through cooling, filter the crystals on a **Buchner funnel**.

9. Dry them and take a melting point, as described in Chapter 12, "The Melting Point Experiment."

GRAVITY FILTRATION

If you find yourself with a flask full of hot solvent, and your product dissolved in it, along with all sorts of trash, this is for you. You'll need more hot solvent, a ringstand with a ring attached, possibly a clay triangle, some filter paper, a clean, dry flask, and a **stemless funnel**. Here's how **gravity filtration** works.

1. Fold up a **filter cone** out of a piece of filter paper (Fig. 57). It should fit nicely, within a single centimeter or so of the top of the funnel. For those who wish to filter with more panache, try using **fluted filter paper** (see "world famous fan-folded fluted filter paper," Fig. 67).
2. Get yourself a **stemless funnel** or, at least, a **short-stemmed funnel**. Why? Go ahead and *use* a stem funnel and watch the crystals come out in the stem as the solution cools, blocking up the funnel (Fig. 58).
3. Put the **filter paper cone** in the **stemless funnel**.
4. Support this in a ring attached to a ringstand (Fig. 59). If the funnel is too small and you think it could fall through the ring, you may be able to get a **wire** or **clay triangle** to support the funnel in the ring (Fig. 60).
5. Put the new, *clean, dry flask* under the funnel to catch the hot solution as it comes through. All set?
6. Get that flask with the solvent, product, and trash hot again. *(No flames!)* You should get some fresh, clean solvent hot as well. *(No flames!)*
7. Carefully pour the hot solution into the funnel. As it is, some solvents evaporate so quickly that product will probably come out on the filter paper. It is often hard to tell the product from the insoluble trash. Then—
8. Wash the filter paper down with *a little hot solvent*. The product will redissolve. The trash won't.
9. You now let the *trash-free* solution cool, and clean crystals should come out. Since you have probably added solvent to the solution, *don't be surprised if no crystals come out of solution. Don't panic either!* Just boil away some of the solvent (not out into the room air, please) let your solution cool, and wait for the crystals again. If they *still* don't come back, just repeat the boiling.

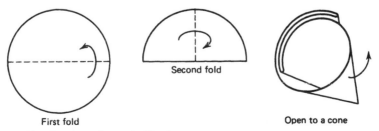

First fold   Second fold   Open to a cone

Fig. 57   Folding filter paper for gravity filtration.

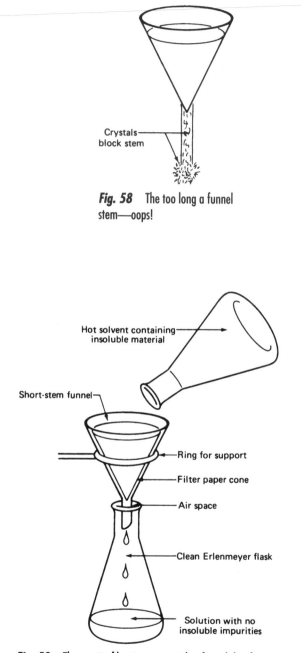

Fig. 58 The too long a funnel stem—oops!

Fig. 59 The gravity filtration setup with a funnel that fits the iron ring.

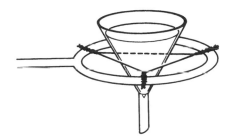

Fig. 60 A wire triangle holding a small funnel in a large iron ring for gravity filtration.

13

Do not boil to dryness!

Somehow, lots of folk think recrystallization means dissolving the solid, then boiling away all the solvent to dryness. No! There must be a way to convince these lost souls that *the impurities will deposit on the crystals.* After the solution has cooled, crystals come out, sit on the bottom of the flask, and *must be covered by solvent!* Enough solvent to keep those nasty impurities dissolved and off the crystals.

THE BUCHNER FUNNEL AND FILTER FLASK

The **Buchner funnel** (Fig. 61) is used primarily for separating crystals of product from the liquid above them. If you have been *boiling your recrystallization solvents dry, you should be horsewhipped* and forced to reread these sections on recrystallization!

1. Get a piece of filter paper large enough to cover all the holes in the bottom plate, yet *not* curl up the sides of the funnel. It is placed *flat* on the plate (Fig. 61).
2. Clamp a **filter flask** to a ringstand. This filter flask, often called a **suction flask**, is a very heavy-walled flask with a sidearm on the neck. A piece of heavy-walled tubing connects this flask to the **water trap** (see Fig. 63).
3. Now use a **rubber stopper** or **filter adapter** to stick the Buchner funnel into the top of the filter flask. The Buchner funnel makes the setup top-heavy and prone to be prone—and broken. Clamp the flask

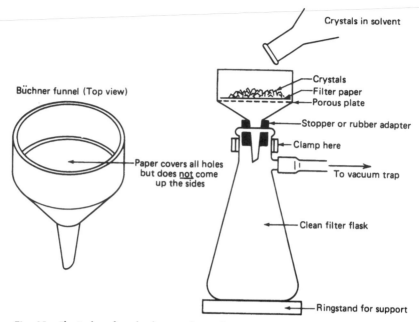

Fig. 61 The Buchner funnel at home and at work.

first, or go get a new Buchner funnel to replace the one you'll otherwise break.

4. The **water trap** is in turn connected to a source of vacuum, most likely, a **water aspirator** (Fig. 62).

5. The faucet on the **water aspirator** should be turned on *full blast*! This should suck down the filter paper, which you now *wet with some of the cold recrystallization solvent*. This will make the paper stick to the plate. You may have to push down on the Buchner funnel a bit to get a good seal between the rubber adapter and the funnel.

6. Swirl and pour the crystals and solvent *slowly, directly into the center of the filter paper*, as if to build a small mound of product there. *Slowly!* Don't flood the funnel by filling it right to the brim and waiting for the level to go down. If you do that, the paper may float up, ruining the whole setup.

7. Use a very small amount of the same cold recrystallization solvent and a spatula to remove any crystals left in the flask. Then you can use some of the *fresh, cold recrystallization solvent* and slowly pour it over the crystals to wash away any old recrystallization solvent and dissolved impurities.

8. Leave the aspirator on and let air pass through the crystals to help them dry. You can put a thin rubber sheet, a **rubber dam**, over the funnel. The vacuum pulls it in, and the crystals are pressed clean and dry. You won't have air or moisture blowing through, and possibly decomposing, your product. Rubber dams are neat.

9. When the crystals are dry, *and you have a* **water trap**, just turn off the water aspirator. Water won't back up into your flask. [If you've been foolhardy and have filtered without a water trap, just remove the rubber tube connected to the filter flask sidearm (Fig. 62).]

10. At this point, you may have a *cake of crystals* in your **Buchner funnel**. The easiest way to handle this is to *carefully lift the cake* of crystals out of the funnel *along with the filter paper*, plop the whole thing onto a larger piece of filter paper, and let the whole thing dry overnight. If you are pressed for time, *scrape the damp filter cake from the filter paper, but don't scrape any filter paper fibers into the crystals*. Repeatedly press the crystals out between dry sheets of filter paper, changing sheets until the crystals no longer show any solvent spot

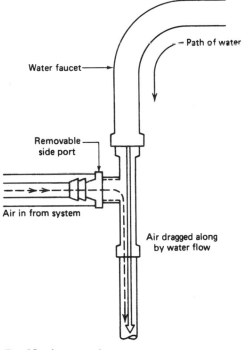

Fig. 62 A water aspirator.

after pressing. Those of you who use **heat lamps** may find your white crystalline product turning into instant charred remains.

11. When your cake is *completely dried*, weigh a vial, put in the product, and weigh the vial again. Subtracting the weight of the vial from the weight of the vial and sample will give the weight of the product. This **weighing by difference** is easier and less messy than weighing the crystals directly on the balance. This weight should be included in the label on your **product vial** (see Chapter 6, "Microscale Jointware" and Chapter 11, "On Products").

Just a Note

I've said that a Buchner funnel is used primarily for separating crystals of product from the liquid above them. And in the section on drying agents, I

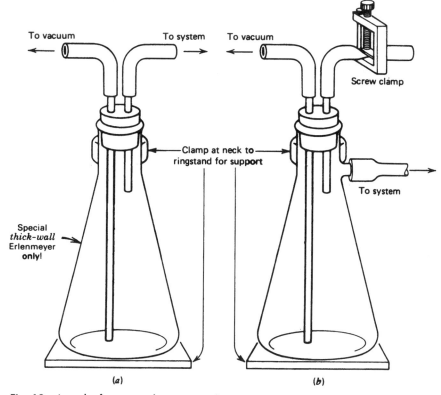

Fig. 63 A couple of water traps hanging around.

tell people to use a gravity filtration setup to separate a drying agent from a liquid product. Recently, I've had some people get the notion that you can Buchner filter products from drying agents. I don't advise that. You will probably lose a lot of your product, especially if it has a low boiling point (<100°C). Under this *vacuum filtration,* your product simply evaporates along with your grade.

13

ACTIVATED CHARCOAL

Activated charcoal is ultrafinely divided carbon with lots of places to suck up big, huge, polar, colored impurity molecules. Unfortunately, if you use too much, it'll suck up **your product**! And, if your product was white or yellow, it'll have a funny gray color from the excess charcoal. Sometimes the impurities are untouched and only the product gets absorbed. Again, it's a matter of trial and error. Try not to use too much. Suppose you've got a *hot solution* of some solid, *and the solution is highly colored.* Well,

1. First, *make sure your product should not **be** colored!*
2. Take the flask with your filthy product off the heat and swirl the flask. This dissipates any superheated areas so that when you add the activated charcoal, the solution doesn't foam out of the flask and onto your shoes.
3. *Add the activated charcoal.* Put a small amount, about the size of a pea, on your spatula; then throw the charcoal in. Stir. The solution should turn black. Stir and heat.
4. Set up the **gravity filtration** and filter off the carbon. It is especially important to *wash off any product caught on the charcoal,* and it is really hard to see anything here. You should take advantage of **fluted filter paper**. It should give a more efficient filtration.
5. Yes, have some extra fresh solvent heated as well. You'll need to add a few milliliters of this to the hot solution to help keep the crystals from coming out on the filter paper. And you'll need more to help wash the crystals off of the paper when they come out on it anyway.
6. This solution should be *much cleaner* than the original solution. If not, you'll have to *add charcoal and filter again.* There is a point of diminishing returns, however, and one or two treatments is usually all you should do. Get some guidance from your instructor.

Your solid products should not be gray. Liquid products (yes, you can do liquids!) will let you know that you didn't get all the charcoal out. Often, you can't see charcoal contamination in liquids while you're working with them. The particles stay suspended for awhile, but after a few days, you can see a layer of charcoal on the bottom of the container. Sneaky, those liquids. By the time the instructor gets to grade all the products—*voila*—the charcoal has appeared.

THE WATER ASPIRATOR: A VACUUM SOURCE

Sometimes you'll need a vacuum for special work like **vacuum distillation** and **vacuum filtration** as with the Buchner funnel. An inexpensive source of vacuum is the **water aspirator** (Fig. 62).

When you turn the water on, the water flow draws air in from the side port on the aspirator. The faster the water goes through, the faster the air is drawn in. Pretty neat, huh? I've shown a plastic aspirator, but many of the older metal varieties are still around.

You may have to pre-test some aspirators before you find one that will work well. It'll depend on the water pressure in the pipes, too. Even the number of people using aspirators on the same water line can affect the performance of these devices. You can test them by going to an aspirator and turning the faucet on *full blast*. It does help to have a sink under the aspirator. If water leaks out the side port, *tell your instructor and find another aspirator*. Wet your finger and place it over the hole in the side port to feel if there is any vacuum. If there is *no vacuum*, tell your instructor and find another aspirator. Some of these old, wheezing aspirators produce a very weak vacuum. You must decide for yourself if the suction is "strong enough." There should be a **splash guard** or rubber tubing leading the water stream directly into the sink. This will keep the water from going all over the room. If you check and don't find such protection, see your instructor. All you have to do with a fully tested and satisfactory aspirator is to hook it up to the **water trap**.

THE WATER TRAP

Every year I run a chem lab, and when someone is doing a **vacuum filtration**, suddenly I'll hear a scream and a moan of anguish, as water backs up into someone's filtration system. Usually there's not much damage, since the filtrate in the suction flask is generally thrown out. For **vacuum distillations**, however, this **suck-back** is disaster. It happens whenever there's a pressure drop on the water line big enough to cause the flow to decrease so that there is a *greater vacuum in the system than in the aspirator*. Water, being water, flows into the system. Disaster.

So, for your own protection, make up a **water trap** from some stoppers, rubber tubing, a thick-walled Erlenmeyer or filter flask, and a screw clamp (Fig. 63). *Do not use garden-variety Erlenmeyers; they may implode without warning.* Two versions are shown. I think the setup using the filter flask is more flexible. The screw clamp allows you to let air into your setup at a controlled rate. You might clamp the water trap to a ringstand when you use it. The connecting hoses have been known to flip unsecured flasks, two out of three times.

WORKING WITH A MIXED-SOLVENT SYSTEM—THE GOOD PART

If, after sufficient agony, you cannot find a single solvent to recrystallize your product from, you may just give up and try a *mixed-solvent system*. Yes, it does mean you mix more than one solvent and *recrystallize using the mixture*. It should only be so easy. Sometimes you are told what the mixture is and the correct proportions. Then it is easy.

For an example, I could use "solvent 1" and "solvent 2," but that's clumsy. So I'll use the ethanol-water system and point out the interesting stuff as I go along.

The Ethanol-Water System

If you look up the solubility of water in ethanol (or ethanol in water), you find an [8]. This means they mix in all proportions. Any amount of one dissolves completely in the other—no matter what. Any volumes, any weights. You name it. The special word for this property is **miscibility**. Miscible solvent systems are the kinds you should use for mixed solvents. They keep you out of trouble. You'll be adding amounts of water to the ethanol, and ethanol to the water. If the two weren't miscible, they might begin to separate and form two layers as you changed the proportions. Then you'd have REAL trouble. So, go ahead. You *can* work with mixtures of solvents that aren't miscible. But don't say you haven't been warned.

The ethanol-water mixture is useful because

1. *At high temperatures, it behaves like alcohol!*
2. *At low temperatures, it behaves like water!*

From this, you should get the idea that it would be good to use a mixed solvent to recrystallize compounds that are *soluble in alcohol* yet *insoluble in water*. You see, each solvent alone cannot be used. If the material is soluble in the alcohol, not many crystals come back from alcohol alone. If the material is insoluble in water, you cannot even begin to dissolve it. So, you have a *mixed solvent*, with the best properties of *both* solvents. To actually perform a *mixed-solvent recrystallization,* you

1. Dissolve the compound in the smallest amount of *hot ethanol*.
2. Add *hot water* until the solution turns cloudy. This **cloudiness** is *tiny crystals of compound coming out of solution*. Heat this solution to dissolve the crystals. If they do not dissolve completely, add *a very little hot ethanol* to force them back into solution.
3. Cool, and collect the crystals on a **Buchner funnel**.

Any solvent pair that behaves the same way can be used. The addition of hot solvents to one another can be tricky. It can be *extremely dangerous* if the boiling points of the solvents are very different. For the *water-methanol mixed solvent*, if 95°C water hits *hot methanol* (B.P. 65.0°C), watch out!

There are other miscible, mixed-solvent pairs, petroleum ether and diethyl ether, methanol and water, and ligroin and diethyl ether among them.

A MIXED-SOLVENT SYSTEM—THE BAD PART

Every silver lining has a cloud. More often than not, compounds "recrystallized" from a mixed-solvent system don't form crystals. Your compound may form an *oil* instead.

Oiling out is what it's called; more work is what it means. Compounds usually oil out if *the boiling point of the recrystallization solvent is higher than the melting point of the compound*, although that's not the only time. In any case, if the oil solidifies, the impurities are trapped in the now solid "oil," and you'll have to purify the solid again.

Don't think you won't ever get oiling out if you stick to single, unmixed solvents. It's just that, with two solvents, there's a greater chance you'll hit on a composition that will cause this.

Temporarily, you can

1. Add more solvent. If it's a mixed-solvent system, try adding more of the solvent the solid is NOT soluble in. Or add more of the OTHER solvent. No contradiction. The point is to *change the composition*. Whether single solvent or mixed solvent, changing the composition is one way out of this mess.
2. Redissolve the oil by heating; then shake up the solution as it cools and begins to oil out. When these smaller droplets finally freeze out, they may form crystals that are relatively pure. They may not. You'll probably have to clean them up again. Just don't use the same recrystallization solvent.

Sometimes, once a solid oils out it doesn't want to solidify at all, and you might not have all day. Try removing a sample of the oil with an eyedropper or a disposable pipette. Then get a glass surface (watch glass) and add a few drops of a solvent that the compound is known to be *insoluble* in (usually water). Then use the rounded end of a glass rod to *triturate the oil with the solvent*. **Trituration** can be described loosely as the beating of an oil into a crystalline solid. Then you can put these crystals back into the rest of the oil. Possibly they'll act as seed crystals and get the rest of the oil to solidify. Again, you'll still have to clean up your compound.

SALTING-OUT

Sometimes you'll have to recrystallize your organic compound from water. No big deal. But sometimes your organic compound is more than ever so slightly insoluble in water, and not all the compound will come back. Solution? Salt solution! A pinch of salt in the water raises the **ionic strength**. There are now charged ions in the water. Some of the water that solvated your compound goes to be with the salt ions. Your organic compound does not particularly like charged ions anyway so more of your organic compound comes out of the solution.

You can dissolve about 36 g of common salt in 100 ml of cold water. That's the upper limit for salt. You can estimate how much salt you'll need to practically saturate the water with salt. Be careful though—if you use too much salt, you may find yourself collecting salt crystals along with your product. (See also the application of salting-out when you have to do an extraction; "Extraction Hints" in chapter 15.)

WORLD FAMOUS FAN-FOLDED FLUTED PAPER

Some training in Origami is *de rigueur* for chemists. It seems that the regular filter paper fold is inefficient, since very little of the paper is exposed. The idea here is to **flute** or **corrugate** the paper, increasing the surface area that is in contact with your filtrate. You'll have to do this several times to get good at it.

Right here let's review the difference between **fold** and **crease**. Folding is folding; creasing is folding, then stomping on it, and running fingers and fingernails over a fold over and over and over. Creasing so weakens the paper, especially near the point, that it may break at an inappropriate time in the filtration.

1. Fold the paper in half, then in half again, then in half again (Fig. 64). Press on this wedge of paper to get the fold lines to stay, but *don't crease. Do this in one direction only.* Either always fold toward you or away from you, but not both.
2. Unfold this cone *twice* so it looks like a semicircle (Fig. 65), and then put it down on a flat surface. Look at it and think for not less than two full minutes the first time you do this.
3. OK. Now try a "fan fold." You alternately fold, first in one direction and then the other, every individual eighth section of the semicircle (Fig. 66).
4. Open the fan and play with it until you get a fairly fluted filter cone (Fig. 67).

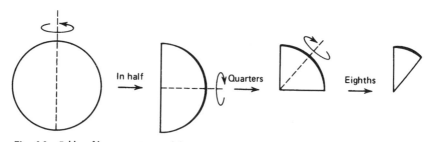

Fig. 64 Folding filter paper into eighths.

5. It'll be a bit difficult, but try to find the two opposing sections that are not folded correctly. Fold them inward (Fig. 67), and you'll have a fantastic fan-folded fluted filter paper of your very own.

P.S. For those with more money than patience, prefolded fan-folded fluted filter paper is available from suppliers.

13

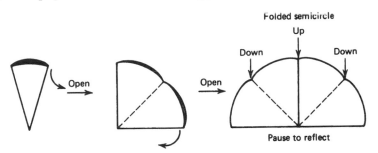

Fig. 65 Unfolding to a sort of bent semicircle.

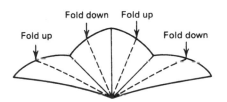

Fig. 66 Refolding to a fan.

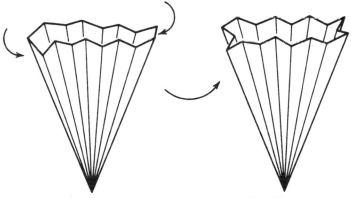

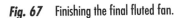

Fig. 67 Finishing the final fluted fan.

RECRYSTALLIZATION: MICROSCALE

Microscale recrystallizations take place in the same way the big ones do. You just use teeny-tiny glassware. Read the general section on crystallization first. Then come back here and:

1. Put your solid in a small test tube.
2. Add a **microboiling stone** or a tiny **magnetic stirring vane** if you're on the stirring hot plate.
3. *Just cover* the solid with recrystallization solvent.
4. Heat this mix in the sand bath; stir with the spinbar or shake the test tube.
5. If the solid isn't dissolving, *consider*:

    a. The stuff not dissolving is insoluble trash; adding lots more solvent won't *even* dissolve it.

    b. The stuff not dissolving is your compound; you need to add 1-2 more drops of solvent.

    This is a decision you'll have to make based on your own observations. Time to take a little responsibility for your actions. I wouldn't, however, add much more than 2–3 ml at the microscale level.
6. OK. Your product has *just dissolved* in the solvent. Add another 10% or so excess of solvent to keep the solid in solution during the hot filtration. If you're going to use *decolorizing carbon*, do it now. Add the carbon; the whole tube should turn black. Heat for 5–10 seconds, and then use Pasteur pipets to filter. You might have to filter twice if you've used carbon.
7. Get your solution into a test tube to cool and recrystallize. Let cool slowly to room temperature first; then put the tube in an ice-water bath for final recrystallization.
8. "I didn't get any crystals." Heat the tube and boil off some of the solvent. **(Caution—Bumping!)** Use a microboiling stone and don't point the tube at anyone. Repeat steps 7 and 8 until you get crystals. The point is to get to *saturation* while hot (all the solid that can be dissolved in the solvent is so), and the crystals will come out when cold.

If you've compared this outline to the large-scale recrystallization, you'll find that, with one exception—keeping a reservoir of hot solvent ready—the only difference here is the size of the equipment. Test tubes for flasks, filter pipets for funnels with paper, and so on. Because you can easily remove solvent at this scale, adding too much solvent is not quite the time-consuming boiling-off you'd have to do at the larger scale.

ISOLATING THE CRYSTALS

First, see the section on "Pipet Filtering—Solids" in Chapter 7 where you remove solvent with a pipet. I also mention using a Hirsch funnel. Just reread the section on the Buchner funnel filtration (Chapter 13, "Recrystallization"), and where you see "Buchner" substitute "Hirsch" and paste copies of the Hirsch funnel drawing over the Buchner funnel. Just realize that the tiny disk of paper can fly up more easily. Think small.

CRAIG TUBE FILTRATION

14

Usually this is called Craig tube crystallization, because you've pipet filtered your hot solution into the bottom of a Craig tube. So if you've recrystallized in something else, re-dissolve the crystals and get this solution into the bottom of a Craig tube (Fig. 68).

1. Carefully put the upper section of the Craig tube into the bottom section. Let the solution slowly cool to room temperature and then put the tube in an ice-water bath.

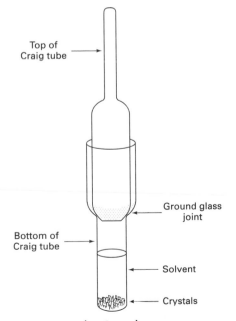

Top of Craig tube

Ground glass joint

Bottom of Craig tube

Solvent

Crystals

Fig. 68 Crystals in Craig tube.

2. Now the fun part. You've got to turn this thing upside down, put it in a centrifuge tube, and *not* have it come apart and spill everything. Easy to say—"Solvent is now removed by inverting the Craig tube assembly into a centrifuge tube..." (Mayo et al., p. 80). Easy to do ... well:

3. Get your centrifuge tube and cut a length of copper wire as long as the tube. Make a loop in the end that's big enough to slip over the end of the Craig tube stem. Twist the wire so that the loop won't open if you pull on it a bit.

4. Bend the loop so that it makes about a right angle from the wire (Fig. 69). Ready?

5. Hold the bottom of the Craig tube between your thumb and forefinger. Put the loop over the end of the tube.

6. Hold the wire gently between your thumb (or forefinger) and the glass tube bottom. Grip the hanging wire and gently pull it downward so that the Craig tube top is pulled into the bottom (Fig. 70). Too much force here and you could snap the stem. Watch it.

7. Once this is adjusted, tightly press the wire to the glass, so that the wire will keep the Craig tube closed when you turn the tube upside down.

8. Put the centrifuge tube upside down over the top of the Craig tube (Fig. 71b). Watch it. If the fit is snug, the copper wire could cause

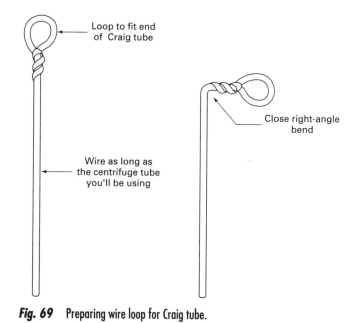

Loop to fit end
of Craig tube

Close right-angle
bend

Wire as long as
the centrifuge tube
you'll be using

Fig. 69 Preparing wire loop for Craig tube.

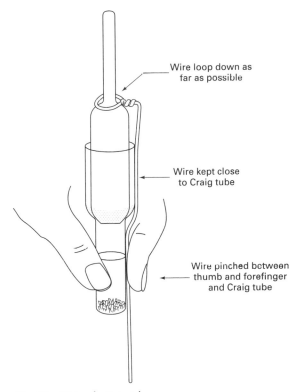

Wire loop down as
far as possible

Wire kept close
to Craig tube

Wire pinched between
thumb and forefinger
and Craig tube

14

Fig. 70 Wiring the Craig tube.

problems, such as breaking the tubes. If the centrifuge tube is too tight, you might consider getting a wider one.

9. Once you get to this point, you may have three situations:

> ***Baby Bear*** (Fig. 71a): The centrifuge tube is a bit shorter than the Craig tube. While not the best, some of the Craig tube peeks out from the centrifuge tube. The centrifuge *might* have clearance in the head to accommodate the extra length. I don't like this. Perhaps you should ask your instructor. Invert the setup and get ready to centrifuge.
>
> ***Mama Bear*** (Fig. 71b): The centrifuge tube is a bit longer than the Craig tube. "A bit" here can vary. I mean you can push the Craig tube up into the centrifuge tube with your fingers so that the Craig tube stem touches the bottom or is within a centimeter or so of the bottom. Push the tube up as far as it will go. Invert the setup and get ready to centrifuge.

Papa Bear (Fig. 71c): The centrifuge tube is a *lot* longer than the Craig tube. You can't push it up to the bottom with your fingers. Grab the copper wire at the end, invert the setup, and quickly *lower* the Craig tube to the bottom of the centrifuge tube. Now you're ready to centrifuge.

CENTRIFUGING THE CRAIG TUBE

Just a few things here:

1. "Where does the centrifuge go when it takes a walk?" The centrifuge must be *balanced*. An even number of tubes, almost identical weights, set in positions opposite each other.
2. If you've cut the copper wire properly, it should stick out only a little bit from the centrifuge tube. You wouldn't want to be flailed by a foot-long

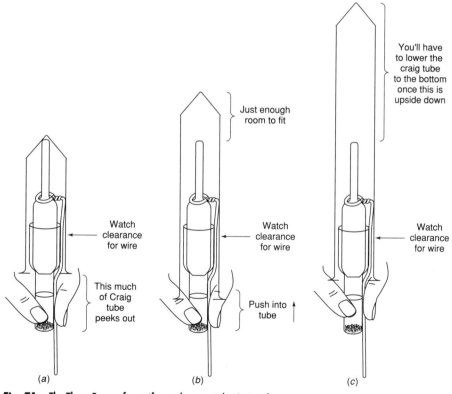

Fig. 71 The Three Bears of centrifuge tubes meet the Craig tube.

copper wire, would you? If necessary, bend it out of the way. Don't bend it so that you lose the wire into the centrifuge tube and can't get it again.

3. If centrifuging one minute is good, centrifuging 15 minutes is *NOT* better. Whatever the directions call for, just do it. Your yield will not improve. Trust me.

Getting the Crystals Out

After you've centrifuged the Craig tube, and the centrifuge has stopped turning

Do not stop the centrifuge with your fingers! Patience!

Pull the centrifuge tubes out, and carefully lift out the Craig tube. The solvent will have filled the bottom of the centrifuge tube, and the crystals will be packed down near the Craig tube top (Fig. 72).

I'd get a glass plate or watch glass, open the Craig tube, and put both sections on the glass to dry. (**CAUTION**—rolling Craig tubes will spread your product on the bench.)

After a bit, you can scrape your crystals into the appropriate tared container. (See "Tare to the Analytical Balance" in Chapter 6, "Microscale Jointware.") Don't try to get all of them; it's usually not worth the effort.

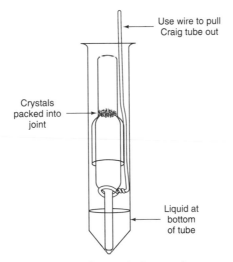

Use wire to pull
Craig tube out

Crystals
packed into
joint

Liquid at
bottom
of tube

Fig. 72 Craig tube crystals after centrifugation.

EXTRACTION
AND WASHING

Extraction is one of the more complex operations you'll do in the organic chemistry lab. For this reason, I'll go over it especially slowly and carefully. Another term you'll see used simultaneously with **extraction** is **washing**. That's because extraction and washing are really the same operation, but each leads to a different end. How else to put this?

Let's make some soup. Put the vegetables, fresh from the store, in a pot. You run cold water in and over them to clean them and throw this water down the drain. Later, you run water in and over them to cook them. You keep this water—it's the soup.

Both operations are similar. Vegetables in a pot in contact with water the first time is a **wash**. You remove unwanted dirt. *You washed with water.* The second time, vegetables in a pot in contact with water is an **extraction**. You've **extracted** essences of the vegetables *into water.* Very similar operations; very different ends.

To put it a little differently,

You would extract good material from an impure matrix.
You would wash the impurities from good material.

The vegetable soup preparation is a *solid-liquid extraction.* So is coffee making. You extract some component(s) of a solid directly into the solvent. You might do a solid-liquid extraction in lab as a separate experiment; *liquid-liquid extractions are routine.* They are so common that if you are told to do an extraction or a washing, *it is assumed*, you will use *two liquids—two insoluble liquids—and a separatory funnel.* The separatory funnel, called a **sep funnel** by those in the know, is a special funnel that you can separate liquids in. You might look at the section on separatory funnels (later in this chapter) right now, and then come back later.

Two insoluble liquids in a separatory funnel will form **layers**; one liquid will float on top of the other. You usually have compounds dissolved in these layers, and either the compound you want is *extracted from one to the other*, or junk you don't want is *washed from one layer to the other.*

Making the soup, you have *no* difficulty deciding what to keep or what to throw away. First you throw the water away; later you keep it. But this can change. In a sep funnel, the layer you want to keep one time may not be the layer you want to keep the next time. Yet, if you throw one layer away prematurely, you are doomed.

NEVER-EVER LAND

Never, never, never, never,
ever throw away any layer, until you are absolutely sure you'll
never need it again. Not very much of your product can be
recovered from the sink trap!

I'm using a word processor, so I can copy this warning over and over again, but let's not get carried away. One more time. Wake Up Out There!

Never, never, never, never,
ever throw away any layer, until you are absolutely sure you'll
never need it again. Not very much of your product can be
recovered from the sink trap!

15

STARTING AN EXTRACTION

To do any extraction, you'll need two liquids or solutions. *They must be insoluble in each other.* **Insoluble** here has a practical definition:

When mixed together, the two liquids form two layers.

One liquid will float on top of the other. A good example is ether and water. Handbooks say that ether is slightly soluble in water. When ether and water are mixed, yes, some of the ether dissolves; most of the ether just floats on top of the water.

Really soluble or *miscible liquid pairs are no good* for extraction and washing. When you mix them, *they will not form two layers*! In fact, they'll *mix in all proportions*. A good example of this is acetone and water. What kinds of problems can this cause? Well, for one, *you cannot perform any extraction with two liquids that are miscible.*

Let's try it. A mixture of, say, some mineral acid (is HCl all right?) and an organic liquid, "Compound A," needs to have that acid washed out of it. You dissolve the compound A-acid mixture in some acetone. It goes into the sep funnel, and you now add water to wash out the acid.

Acetone is miscible in water. No layers form! You lose!

Back to the lab bench. Empty the funnel. Start over. This time, having called yourself several colorful names because you should have read this section thoroughly in the first place, you dissolve the Compound A–acid mixture in ether and put it into the sep funnel. Add water, and *two layers form*! Now you can wash the acid from the organic layer to the water layer. The water layer can be thrown away.

Note that the acid went into the water, *then the water was thrown out*! So we call this a **wash**. If the water layer had been saved, we'd say the acid had been **extracted into the water layer**. It may not make sense, but that's how it is.

Review:

1. You *must have two insoluble liquid layers* to perform an extraction.
2. *Solids must be dissolved in a solvent*, and that solvent must be insoluble in the other extracting or washing liquid.
3. If you are washing or extracting an organic liquid, dissolve it into another liquid, *just like a solid*, before extracting or washing it.

So these terms, **extraction and washing** are related. Here are a few examples.

1. Extract with ether. Throw ether together with the solution of product and pull out *only the product into the ether*.
2. Wash with 10% NaOH. Throw 10% NaOH together with the solution of product and pull out *everything but product into the NaOH*.
3. You can even extract with 10% NaOH.
4. You can even wash with ether.

> So extraction is pulling out what you want from all else!
> Washing is pulling out all else from what you want.

And please note—*you always do the pulling from one layer into another*. That's also *two immiscible liquids*.

You'll actually have to do a few of these things before you get the hang of it, but bear with me. When your head stops hurting, reread this section.

DUTCH UNCLE ADVICE

Just before I go on to the separatory funnel, I'd like to comment on a few questions I keep hearing when people do washings and extractions.

1. **"Which layer is the water layer?"** Look at both layers in the funnel and get an idea of how big they are in relation to one another. Now *add water to the funnel*. Watch where the water goes. Watch which layer grows. Water to water. That's how you find the water (aqueous) layer. Don't rely on odor or color. Enough ether dissolves in water to give the water layer the odor of an ether layer; just enough of a highly colored substance in the wrong layer can mislead you.

2. **"How come I got three layers?"** Sometimes, when you pour fresh water or some other solvent into the funnel, you get a small amount hanging at the top, and it looks like there are three different layers. Yes, it *looks* as if there are three different layers, but there are not three different layers. Only two layers, where part of one has lost its way. Usually, this mysterious third layer looks just like its parent, and you might gently swirl the funnel and its contents to reunite the family.

3. **"What's the density of sodium hydroxide?"** You've just done a wash with 5–10% sodium hydroxide solution, you've just read something about finding various layers in the funnel by their densities, and, by this question, you've just shown that you've missed the point. Most wash solutions are 5–10% active ingredient dissolved in water. This means they are 90–95% water. Looking up the density of the solid reagents, then, is a waste of time. The density of these solutions is *very close* to that of water. (10% NaOH has a specific gravity of 1.1089.)

4. **"I've washed this organic compound six times with sodium bicarbonate solution, so why's it not basic yet?"** This involves finding the pH of the organic layer. I'll give it away right now. *You cannot find the pH of an organic layer.* Not directly. You find the pH of the aqueous layer that's been in contact with the organic layer. If the aqueous layer is on the top, dip a glass rod into it and touch the glass rod to your test paper. If the aqueous layer is on the bottom and your sep funnel is in a ring, let a drop of the aqueous layer out of the funnel to hand on the outlet tip. Transfer the drop to your test paper. Warning. Be *sure* you are testing the aqueous layer. Some organics are very tenacious and can get onto your glass rod. The organic layer may wet the test paper, but without water any color you see doesn't mean much.

THE SEPARATORY FUNNEL

Before going on to some practical examples, you might want to know more about where all this washing and extracting is carried out. I've mentioned that it's a special funnel called a **separatory funnel** (Fig. 73) and that you can impress your friends by calling it a **sep funnel**. Here are a few things you should know.

The Stopper

At the top of the sep funnel is a 𝕋 glass stopper. There is a number, commonly 𝕋 22, possibly 𝕋 19/22, on the stopper head. *Make sure this number is on the head and that it is the same as the number marked on the funnel.* If this stopper is not so marked, you may find the product leaking over your shoes when you turn the sep funnel upside down. Try *not to grease this stopper* unless you plan

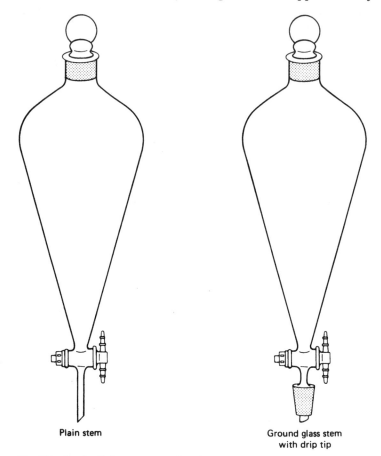

Plain stem

Ground glass stem
with drip tip

Fig. 73 Garden Variety separatory funnels.

to sauté your product. Unfortunately, these stoppers tend to get stuck in the funnel. The way out is to be sure you don't get the ground glass surfaces wet with product. How? Pour solutions into the sep funnel as carefully as you might empty a shotglass of Scotch into the soda. Maybe *use a funnel.* To confuse matters, I'll suggest you use a light coating of grease. Unfortunately, my idea of light and your idea of light may be different.

Consult your instructor!

The Glass Stopcock

This is the time-honored favorite of separatory funnel makers everywhere. There is a notch at the small end that contains either a rubber ring *or* a metal clip, *but not both*! There are two purposes for the ring.

1. To keep the stopcock from falling out entirely. Unfortunately, the rubber rings are not aware of this, and the stopcock falls out anyway.
2. To provide a sideways pressure, pulling the stopcock in, so that it will not leak. Names and addresses of individuals whose stopcocks could not possibly leak and did so anyway will be provided on request. So provide a little sideways pressure of your own.

When you grease a glass stopcock (and you must), do it very carefully so that the film of grease does *not* spread into the area of the stopper that comes in contact with any of your compound (Fig. 74).

The Teflon Stopcock

In wide use today, the **Teflon stopcock** (Fig. 75) requires no grease and will not freeze up! The glass surrounding the stopcock *is not ground glass and cannot be used in funnels that require ground glass stopcocks*! The Teflon stopcocks are infinitely easier to take care of. A Teflon washer, a rubber ring, and, finally, a Teflon nut, are placed on the threads of the stopcock. This nut holds the whole thing on. Any leakage at this stopcock results from

1. A loose Teflon nut. Tighten it.
2. A missing Teflon washer or rubber ring. Have it replaced.
3. An attempt to place the wrong size or taper Teflon stopcock into the funnel. This is extremely rare. Get a new funnel.

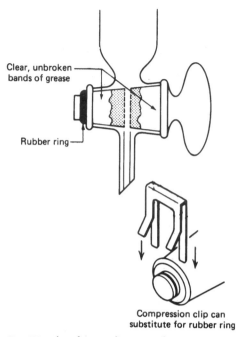

Clear, unbroken
bands of grease

Rubber ring

Compression clip can
substitute for rubber ring

Fig. 74 The infamous Glass Stopcock.

Emergency stopcock warning!

Teflon may not stick, *but it sure can flow*! If the stopcock is extremely tight, the Teflon will bond itself to all the nooks and crannies in the glass in interesting ways. When you're through, always loosen the Teflon nut and "pop the stopcock" by pulling on the handle. The stopcock should be loose enough to spin freely when spun with one finger—*then remember to tighten it again before you use it*.

It seems to me that I'm the only one who reads the little plastic bags that hold the stopcock parts. Right on the bags it shows that after the stopcock goes in, *the Teflon washer goes on the stem first*, followed by the rubber ring, and then the Teflon nut (Fig. 75). So why do I find most of these things put together incorrectly?

THE STEM

The stem on a sep funnel can either be straight or have a ground glass joint on the end (Fig. 73). The ground glass joint fits the other jointware you may have and can be used that way as an **addition funnel** to add liquids or solutions into a setup (see "Addition and Reflux" in Chapter 22). You can use this type of separatory funnel as a sep funnel. You can't, however, use the straight-stem separatory funnel as an addition funnel without some help; remember, straight glass tubes don't fit ground glass joints (see "The Adapter with Lots of Names").

WASHING AND EXTRACTING VARIOUS THINGS

Now, getting back to extractions, only four classes of compounds are commonly handled in undergraduate extractions or washings.

1. **Strong acids.** The mineral acids and organic acids (e.g., benzoic acid). You usually *extract these into sodium bicarbonate solution or wash them with it.*
2. **Really weak acids.** Usually phenols or substituted phenols. Here you'd use a *sodium hydroxide solution for washing or extraction.* You need a *strong base* to work with these weak acids.
3. **Organic bases.** Any organic amine (aniline, triethylamine, etc.). As you use bases to work with acids, use a **dilute acid** (5–10% HCl, say) to extract or wash these bases.
4. **Neutral compounds.** All else, by these definitions (e.g., amides, ethers, alcohols, hydrocarbons).

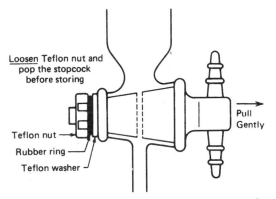

Loosen Teflon nut and
pop the stopcock
before storing

Pull
Gently

Teflon nut

Rubber ring

Teflon washer

Fig. 75 Extreme close-up of teflon stopcock popping ritual.

HOW TO EXTRACT AND WASH WHAT

Here are some practical examples of washings and extractions, covering various types and mixtures and separations and broken down into the four classifications listed previously.

1. ***A strong organic acid.*** Extract into sat'd (saturated) sodium bicarbonate solution.
    (**Caution!** *Foaming and fizzing and splitting and all sorts of carrying on.*) The weak base turns the strong acid into a salt, and the salt dissolves in the water-bicarbonate solution. Because of all the fizzing, you'll have to be very careful. Pressure can build up and blow the stopper out of the funnel. Invert the funnel. *Point the stem away from everyone up and toward the back of the hood*—and open the stopcock to vent or "burp" the funnel.

    a. *To recover the acid*, add conc. (concentrated) HCl until the solution is acidic. Use pH or litmus paper to make sure. Yes, the solution really fizzes and bubbles. You should use a large beaker so material isn't thrown onto the floor if there's too much foam.

    b. *To wash out the strong acid*, just throw the solution of bicarbonate away.

2. ***A weakly acidic organic acid.*** Extract into 10% NaOH–water solution. The strong base is needed to rip the protons out of weak acids (they don't want to give them up) and turn them into salts. Then they'll go into the NaOH-water layer.

    a. *To recover the acid*, add conc. HCl until the solution of base is acid when tested with pH or litmus paper.

    b. *To wash out the weak acid*, just throw this NaOH–water solution away.

3. ***An organic base.*** Extract with 10% HCl–water solution. The strong acid turns the base into a salt. (This *turning the whatever into a salt that dissolves in the water solution* should be pretty familiar to you by now. Think about it.). Then the salt goes into the water layer.

    a. *To recover the base*, add ammonium hydroxide to the water solution until the solution is *basic* to pH or litmus paper. *Note that this is the reverse of the treatment given to organic acids.*

b. *To wash out an organic base, or any base,* wash as previously noted and throw out the solution.

4. *A neutral organic.* If you've extracted strong acids first, then weak acids, and then bases, there are only neutral compound(s) left. If possible, just remove the solvent that now contains *only your neutral compound.* If you have *more than one neutral compound,* you may want to extract one from the other(s). You'll have to find *two different immiscible organic liquids,* and *one liquid must dissolve only the neutral organic compound you want!* A tall order. You must count on *one neutral organic compound being more soluble in one layer than in the other.* Usually the separation is *not clean—not complete.* And you have to do more work.

What's "more work"? That depends on the results of your extraction.

The Road to Recovery—Back-Extraction

I've mentioned **recovery** of the four types of extractable materials, but that's not all the work you'll have to do to get the compounds in shape for further use.

1. If the recovered material is *soluble in the aqueous recovery solution,* you'll have to do a **back-extraction**.

   a. Find a solvent that dissolves your compound and is *not miscible in the aqueous recovery solution.* This solvent should boil at a low temperature (<100°C), since you will have to remove it. Ethyl ether is a common choice. (**Hazard!** *Very flammable*).

   b. *Then you extract your compound back from the aqueous recovery solution into this organic solvent.*

   c. Dry this organic solution with a drying agent (see Chapter 10, "Drying Agents").

   d. Now you can remove the organic solvent. Either distill the mixture or evaporate it, perhaps on a steam bath. All this is done away from flames and in a hood.

When you're through removing the solvent and you're product is not pure, clean it up. If your product is a liquid, you might distill it; if a solid, you might recrystallize it. Make sure it is clean.

2. If the recovered material is *insoluble in the aqueous recovery solution* and *it is a solid,* collect the crystals on a Buchner funnel. If they are *not pure,* you should recrystallize them.

3. If the recovered material is *insoluble in the aqueous recovery solution* and *it is a liquid,* you can use your separatory funnel to *separate the aqueous recovery solution from your liquid product. Then dry your liquid product and distill it if it is not clean. Or you might just do a back-extraction* as just described. This has the added advantage of getting out the small amount of liquid product that dissolves in the aqueous recovery solution and increases your yield. Remember to dry the back-extracted solution before you remove the organic solvent. Then distill your liquid compound if it is not clean.

A SAMPLE EXTRACTION

I think the only way I can bring this out is to use a typical example. This may ruin a few lab quizzes, but if it helps, it helps.

Say you have to separate a mixture of *benzoic acid (1), phenol (2), p-toluidine (4-methylaniline) (3),* and *anisole (methoxybenzene) (4)* by extraction. The numbers refer to the class of compound, as previously listed. We're assuming that none of the compounds react with any of the others and that you know we're using all four types as indicated. Phenol and 4-methylanaline are corrosive toxic poisons, and if you get near these compounds in lab, *be very careful.* When they are used as an example on these pages, however, you are quite safe. Here's a sequence of tactics.

1. Dissolve the mixture in ether. Ether is insoluble in the water solutions you will extract into. Ether happens to dissolve all four compounds. Aren't you lucky? You bet! It takes lots of hard work to come up with the "typical student example."

2. Extract the ether solution with 10% HCl. This converts *only* compound 3, the basic p-toluidine, into the hydrochloride salt, which dissolves in the 10% HCl layer. You have just *extracted a base with an acid solution.* Save this solution for later.

3. Now extract the ether solution with sat'd sodium bicarbonate solution. *Careful!* Boy will this fizz! Remember to swirl the contents and release the pressure. The weak base converts *only* compound 1, the

benzoic acid, to a salt, which dissolves in the sat'd bicarbonate solution. Save this for later.

4. Now extract the ether solution with the 10% NaOH solution. This converts the compound 2, *weak acid*, phenol, to a salt, which dissolves in the 10% NaOH layer. Save this for later. *If you do this step before step 3; that is, extract with 10% NaOH solution before the sodium bicarbonate solution, both the weak acid,* phenol, *and the strong acid,* benzoic acid, *will be pulled out into the sodium hydroxide.* Ha-Ha. This is the usual kicker they put in lab quizzes, and people always forget it.

5. The only thing left is the neutral organic compound dissolved in the ether. Just drain this into a flask.

So, now we have four flasks with four solutions with one component in each. *They are separated.* You may ask, "How do we get these back?"

1. *The basic compound (3).* Add ammonium hydroxide until the solution turns basic (test with litmus or pH paper). The p-toluidine, or organic base (3), is regenerated.

2. *The strong acid or the weak acid (1,2).* A bonus. Add dilute HCl until the solution turns acidic to an indicator paper. Do it to the other solution. Both acids will be regenerated.

3. *The neutral compound (4).* It's in the ether. If you evaporate the ether *(No flames!),* the compound should come back.

Now, when you recover these compounds, sometimes they don't come back in such good shape. You will have to do more work.

1. Addition of HCl to the benzoic acid extract will produce huge amounts of white crystals. Get out the Buchner funnel and have a field day! Collect all you want. But they won't be in the best of shape. Recrystallize them. (**Note:** *This compound is insoluble in the aqueous recovery solution.*)

2. The phenol extract is a different matter. You see, *phenol is soluble in water,* and it doesn't come back well at all. So, get some fresh ether, extract the phenol from HCl solution to the ether, and evaporate the ether. Sounds crazy, no? No. Remember, I called this a **back-extraction**, and you'll have to do this more often than you would like to believe. (**Note:** *This compound is soluble in the aqueous recovery solution.*)

3. The p-toluidine should return after the addition of ammonium hydroxide. Recrystallize it from ethanol so that it also looks respectable again.

4. The neutral anisole happens to be a *liquid (B.P. 155°C)*, and you'll have to take care when you evaporate the ether so you don't lose much anisole. Of course, you shouldn't expect to see any crystals. Now this neutral anisole liquid that comes back after you've evaporated the ether *(no flames!)* will probably be contaminated with a little bit of all the other compounds that started out in the ether. You will have to purify this liquid, probably by a simple distillation.

   You may or may not have to do all of this with the other solutions or with any other solution you ever extract in your life. You must choose. Art over science. As confusing as this is, I have simplified it a lot. Usually, you have to extract these solutions more than once, and the separation is not as clean as you'd like. Not 100% but pretty good. If you are still confused, see your instructor.

PERFORMING AN EXTRACTION OR WASHING

1. Suspend a sep funnel in an iron ring.
2. Remove the stopper.
3. *Make sure the stopcock is closed!* You don't really want to scrape your product off the benchtop.
4. Add the solution to be extracted or washed. Less than half full, please. Add the extraction or washing solvent. An equal volume is usually enough. The funnel is funnel shaped, and the equal volumes won't look equal.
5. Replace the stopper.
6. Remove the sep funnel from the iron ring. Hold stopper and stopcock tightly. Pressure may build up during the next step and blow your product out onto the floor.
7. Invert the sep funnel (Fig. 76).

   **Point the stem up away from everyone—
   up into the back of a hood if at all possible!**

   Make *sure* the liquid has drained down away from the stopcock, then *slowly* open the stopcock. You may hear a woosh, possibly a pffffft, as

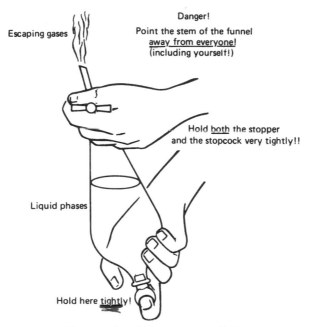

Danger!
Point the stem of the funnel
<u>away from everyone!</u>
(including yourself!)

Escaping gases

Hold <u>both</u> the stopper
and the stopcock very tightly!!

Liquid phases

Hold here <u>tightly</u>!

Fig. 76 Holding a sep funnel so as not to get stuff all over.

the pressure is released. This is due to the high vapor pressure of some solvents or to a gas evolved from a reaction during the mixing. This can cause big trouble when you are told to neutralize acid, by washing with sodium carbonate or sodium bicarbonate solutions.

8. Close the stopcock!
9. Shake the funnel gently, invert it, and open the stopcock again.
10. Repeat steps 8 and 9 until no more gas escapes.
11. If you see that you might get an **emulsion**—*a fog of particles*—with this gentle inversion, *do NOT shake the funnel vigorously.* You might have to continue the rocking and inverting motions 30 to 100 times, as needed, to get a separation. Check with your instructor and with the hints on breaking up emulsions (see "Extraction Hints," following). Otherwise, shake the funnel vigorously about 10 times, in order to get good distribution of the solvents and solutes. Really shake it.
12. Put the sep funnel back in the iron ring.
13. *Remove the glass stopper.* Otherwise the funnel won't drain, and you'll waste your time just standing there.
14. Open the stopcock and let the bottom layer drain off into a flask.

15. Close the stopcock, swirl the funnel gently, and then wait to see if any more of the bottom layer forms. If so, collect it. If not, assume you got it all in the flask.
16. Let the remaining layer out into another flask.

To extract any layer again, return that layer to the sep funnel, *add fresh extraction or washing solvent*, and repeat this procedure starting from step 5.

Never, never, never, never,
ever throw away any layer, until you are absolutely sure you'll
never need it again. Not very much of your product can be
recovered from the sink trap!

EXTRACTION HINTS

1. Several smaller washings or extractions are better than one big one.
2. Extracting or washing a layer *twice*, perhaps three times, is usually enough. Diminishing returns set in after that.
3. Sometimes you'll have to find out which layer is **the water layer**. This is so simple, it confounds everyone. Add 2–4 drops of each layer to a test tube containing 1 ml of water. Shake the tube. If the stuff *doesn't dissolve* in the water, it's *not* an aqueous (water) layer. The stuff may sink to the bottom, float on the top, do *both*, or even *turn the water cloudy*! It will *not*, however, dissolve.
4. If *only the top layer* is being extracted or washed, *it does not have to be removed from the funnel*, ever. Just drain off the bottom layer, and then add more *fresh* extraction or washing solvent. Ask your instructor about this.
5. You *can* combine the extracts of a multiple extraction, *if they have the same material in them*.
6. If you have to wash your organic compound with water, and the organic is *slightly soluble in water*, try washing with *saturated salt solution*. The theory is that if all that salt dissolved in the water, what room is there for your organic product? This point is a favorite of quizmakers and should be remembered. It's the same thing that happens when you add salt to reduce the solubility of your compound during a crystallization (see "Salting-Out" earlier in this chapter).

7. If you get an **emulsion**, you have not two distinct layers, but a kind of a *fog of particles*. Sometimes you can break up the charge on the suspended droplets by adding a little salt, or some acid or base. Or add ethanol. Or stir the solutions slowly with a glass rod. Or gravity filter the entire contents of your separatory funnel through filter paper. Or laugh. Or cry. Emulsion-breaking is a bit of an art. Careful with the acids and bases, though. They can react with your product and destroy it.

8. If you decide to add salt to a sep funnel, don't add so much that it clogs up the stopcock! For the same reason, keep drying agents out of sep funnels.

9. Sometimes some material comes out, or it will not dissolve in the two liquid layers, and hangs in there in the **interface**. It may be that there's not enough liquid to dissolve this material. One cure is to *add more fresh solvent of one layer or the other*. The solid may dissolve. If there's no room to add more, you may have to remove *both* (yes, both) layers from the funnel, and try to dissolve this solid in either of the solvents. It can be confusing. If the material does *not* redissolve, then it is a new compound and should be saved for analysis. You should see your instructor for that one.

EXTRACTION AND WASHING: MICROSCALE

16

Instead of a sep funnel, you use a conical vial and some Pasteur pipets. First you mix your extraction solvent with your product; then you separate the two liquids.

MIXING

1. Get the material to be extracted into an appropriately sized conical vial. This vial should be at *minimum* twice the volume of the liquid you want to extract. Usually, this conical vial is the reaction vial for the experiment, so the choice is easily made (forced on you).
2. Add an appropriate solvent (**not**—repeat—**not** extracting liquid) so the volume of the mixture is about 1 ml. (What if it's about 1 ml already?)
3. Add about 1 ml of extracting solvent to the vial.
4. At this point, cap the vial and shake to mix (**Caution**—Leaks!). Better you should add a **magnetic spinning vane** and spin the two layers for a minute or so.

SEPARATION: REMOVING THE BOTTOM LAYER (FIG. 77)

1. Have an empty vial ready.
2. Pre-wet a Pasteur pipet and bulb combo with the extraction solvent. (See Chapter 7, "Pipet Tips.")
3. Squeeze the bulb.
4. Stick the pipet into the vial down to the bottom.
5. Slowly—read that word again—slowly, draw the bottom layer up into the pipet. Get it all now. (Actually, you could do this a few times if necessary.) And don't suck any liquid **into** the rubber bulb!
6. Squirt the bottom layer into an empty vial.

SEPARATION: REMOVING THE TOP LAYER (FIG. 78)

1. Have an empty vial ready.
2. Pre-wet a Pasteur pipet and bulb combo with the extraction solvent. (See Chapter 7, "Pipet Tips.")
3. Really squeeze the bulb.

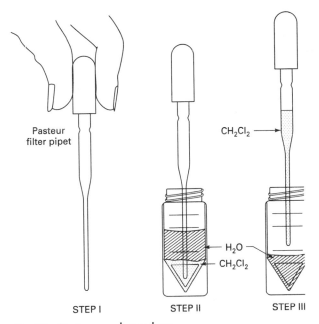

Pasteur
filter pipet

CH$_2$Cl$_2$

H$_2$O

CH$_2$Cl$_2$

STEP I   STEP II   STEP III

Fig. 77 Pipet removes bottom layer.

16

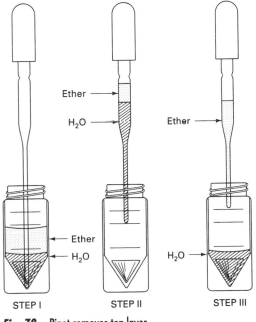

Ether

H$_2$O

Ether

Ether

H$_2$O

H$_2$O

STEP I   STEP II   STEP III

Fig. 78 Pipet removes top layer.

4. Stick the pipet into the vial down to the bottom.

5. Slowly—read that word again—slowly, draw **all**—yes **all**—both layers—into the pipet. **Don't slurp!** If you draw lots of air into the pipet *after* both layers are in there, the air bubble will mix both layers in the pipet. Even worse, because of evaporation into this new "fresh" air, vapor pressure buildup can make both layers fly out back into the vial, onto the benchtop, on your hands, wherever. And don't suck any of this **into** the rubber bulb!

6. Squeeze the bulb to put the *bottom* layer back into the original vial.

7. Squirt the top layer into an empty vial.

AND NOW—BOILING STONES

17

All you want to do is start a distillation. Instructor walks up and says,

"Use a boiling stone or it'll bump."
"But I'm only gonna...."
"Use a boiling stone or it'll bump."
"It's started already and ..."
"Use a boiling stone or it'll bump."
"I'm not gonna go and..."

Suddenly—Woosh! Product all over the bench! Instant failure. Next time you put a boiling stone in before you start. No bumping. But your instructor won't let you forget the time you did it your way.

Don't let this happen to you. Use a **brand-new boiling stone** every time you have to boil a liquid. A close-up comparison between a boiling stone and the inner walls of a typical glass vessel reveals thousands of tiny nucleating points on the stone where vaporization can take place, in contrast to the smooth glass surface that can hide unsightly hot spots and lead to bumping, a massive instantaneous vaporization that will throw your product all over.

Caution! Introduction of a boiling stone into hot liquid may result in instant vaporization and loss of product. Remove the heat source, swirl the liquid to remove hot spots, then add the boiling stone.

Used as directed, the boiling stone will relieve minor hot spots and prevent loss of product through bumping. So remember...*whenever you boil, wherever you boil.*

Always use a fresh boiling stone!

Don't be the last on your bench to get this miracle of modern science made exclusively from nature's most common elements.

SOURCES OF HEAT

18

Many times you'll have to heat something. Don't just reach for the Bunsen burner. That flame you start just may be your own. There are alternative sources you should think of *first*.

THE STEAM BATH

If one of the components boils below 70°C and you use a *Bunsen burner*, you may have a hard time putting out the fire. Use a **steam bath**!

1. Find a steam tap. It's like a water tap, only this one dispenses steam. (**Caution!** *You can get burned.*)
2. Connect tubing to the tap *now*. It's going to get awfully hot in use. Make sure you've connected a piece that'll be long enough to reach your steam bath.
3. Don't connect this tube to the steam bath yet! Just put it into a sink. Because steam lines are usually full of water from condensed steam, *drain the lines first*; otherwise you'll waterlog your steam bath.
4. *Caution!* Slowly open the steam tap. You'll probably hear bonking and clanging as steam enters the line. Water will come out. It'll get hotter and may start to spit.
5. Wait until the line is mostly clear of water, then turn off the steam tap. *Wait for the tubing to cool.*
6. Slowly, carefully, and cautiously, *making sure the tube is not hot*, connect the tube to the inlet of the steam bath. This is the *uppermost* connection on the steam bath.
7. Connect another tube to the outlet of the steam bath—the *lower* connection—and to a drain. Any water that condenses in the bath while you're using it will drain out.

Usually, steam baths have concentric rings as covers. You can control the "size" of the bath by removing various rings.

Never do this after you've started the steam. You will get burned!

And don't forget—round-bottom flasks should be about halfway in the bath. You shouldn't really let steam rise up all around the flask. Lots of steam will certainly steam up the lab and may expose you to corrosion inhibitors (morpholine) in the steam lines (Fig. 79).

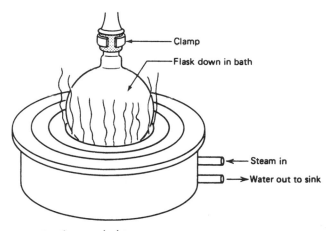

Fig. 79 The steam bath in use.

THE BUNSEN BURNER

The first time you get the urge to take out a Bunsen burner and light it up, *don't*. You may blow yourself up. Please check with your instructor to see if you even need a burner. Once you find out that you *can* use a burner, assume that the person who used it last didn't know much about burners and take some precautions so as not to burn your eyebrows off.

Now Bunsen burners are not the only kind. There are **Tirrill burners** and **Mcker burners** as well. Some are fancier than others, but they work pretty much the same. So when I say **burner** anywhere in the text it could be any of them.

1. Find the **needle valve.** This is at the base of the burner. Turn it fully clockwise (inward) to stop the flow of gas completely. If your burner doesn't have a needle valve, it's a traditional Bunsen burner and the gas flow has to be regulated at the bench stopcock (Fig. 80). This can be dangerous, especially if you have to reach over your apparatus and burner to turn off the gas. Try to get a different model.
2. There is a **movable collar** at the base of the burner which controls air flow. For now, see that *all* the holes are closed (i.e., no air gets in).
3. Connect the burner to the bench stopcock by some tubing and turn the bench valve *full on*. The bench valve handle should be parallel to the outlet (Fig. 80).

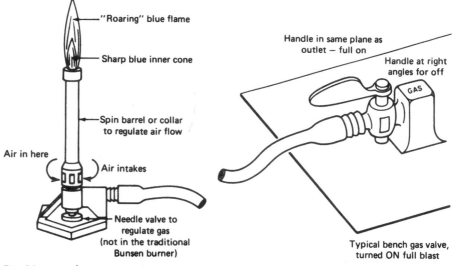

Air in here

Fig. 80 More than you care to know about burners.

4. Now, *slowly* open the needle valve. You may be just able to hear some gas escaping. Light the burner. *Mind your face!* Don't look down at the burner as you open the valve.

5. You'll get a wavy yellow flame, something you don't really want. But at least it'll light. Now open the air collar a little. The yellow disappears; a blue flame forms. This is what you want.

6. Now, adjust the needle valve and collar (the adjustments play off each other) for a steady blue flame.

Burner Hints

1. Air does not burn. You must wait until the gas has pushed the air out of the connecting tubing. Otherwise, you might conclude that none of the burners in the lab work. Patience, please.

2. When you set up the distillation or reflux, don't waste a lot of time *raising* and *lowering* the entire setup so that the burner will fit. This is nonsense. Move the burner! Tilt it! (Fig. 81). If you leave the burner motionless under the flask, you may scorch the compound and your precious product can become a "dark intractable material."

3. Placing a wire gauze between the flame and the flask spreads out the heat evenly. Even so, the burner may have to be moved around. Hot

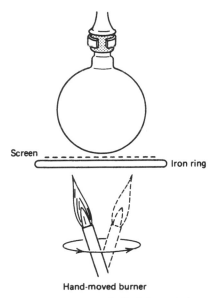

Screen

Iron ring

Hand-moved burner

Fig. 81 Don't raise the flask, lower the burner.

spots can cause star cracks to appear in the flask (See "Round-Bottom Flasks" in Chapter 4, "Generic Jointware").

4. *Never* place the flask in the ring without a screen (Fig. 82). The iron ring heats up faster than the flask, and the flask cracks in the nicest line around it you've ever seen. The bottom falls off, and the material is all over your shoes.

THE HEATING MANTLE

A very nice source of heat, the heating mantle takes some special equipment and finesse.

1. ***Variable voltage transformer.*** The transformer takes the quite lethal 120 V from the wall socket and can change it to an equally dangerous 0 to 120 V, depending on the setting on the dial. Unlike temperature settings on a Mel-Temp, on a transformer 0 means 0 V, 20 means 20 V, and so on. I like to start at 0 V and work my way up. Depending on how much heat you want, values from 40 to 70 seem to be good places to start. Also, you'll need a cord that can plug into both the transformer and the heating mantle.

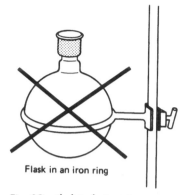

Flask in an iron ring

Fig. 82 Flask in the iron ring.

2. ***Traditional fiberglass heating mantle.*** An electric heater wrapped in fiberglass insulation and cloth that looks vaguely like a catcher's mitt (Fig. 83).

3. ***Thermowell heating mantle.*** You can think of the Thermowell heating mantle as the fiberglass heating mantle in a can. In addition, there is a hard ceramic shell that your flask fits into (Fig. 84). Besides just being

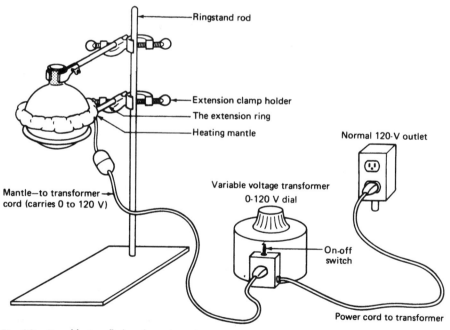

Ringstand rod

Extension clamp holder
The extension ring
Heating mantle

Normal 120-V outlet

Mantle—to transformer
cord (carries 0 to 120 V)

Variable voltage transformer
0-120 V dial

On-off
switch

Power cord to transformer

Fig. 83 Round-bottom flask and mantle ready to go.

more mechanically sound, it'll help stop corrosive liquids from damaging the heating element if your flask cracks while you're heating it.

4. ***Things not to do.***

    a. *Don't ever plug the mantle directly onto the wall socket!* I know, the curved prongs on the mantle connection won't fit, but the straight prongs on the adapter cord will. Always use a variable voltage transformer and start with the transformer set to zero.

    b. *Don't use too small a mantle.* The only cure for this is to *get one that fits properly.* The poor contact between the mantle and the glass doesn't transfer heat readily, and the mantle burns out.

    c. *Don't use too large a mantle.* The only good cure for this is to *get one that fits properly.* An acceptable fix is to *fill the mantle with sand, after the flask is in,* but before you turn the voltage on. Otherwise, the mantle will burn out.

Hint. When you set up a heating mantle to heat any flask, usually for **distillation** or **reflux**, put the mantle on an iron ring and keep it clamped a few inches above the desktop (Fig. 83). Then clamp the flask *at the neck*, in case you have to remove the heat quickly. You can just unscrew the lower clamp and drop the mantle and iron ring.

PROPORTIONAL HEATERS AND STEPLESS CONTROLLERS

In all these cases of heating liquids for distillation or reflux, we really control the *electric power*, not the heat or temperature directly. Power is applied to the heating elements, and they warm up. Yet the final temperature is determined by the heat loss to the room, the air, and, most important, the flask you're heating. There are several types of electric power controls.

Hard ceramic surface

Metal can

Heating element inside can

To power controller

Fig. 84   A Thermowell heating mantle.

1. ***The variable voltage transformer.*** We've discussed this just previously. Let me briefly restate the case: Set the transformer to 50 on the 0–100 dial and you get 50% of the line voltage, all the time, night and day, rain or shine.

2. ***The mechanical stepless controller.*** This appears to be *the* inexpensive replacement for the variable voltage transformer. Inside one model there's a small heating wire wound around a bimetal strip with a magnet at one end (Fig. 85). A plunger connected to the dial on the front panel changes the distance between the magnet and a metal plate. With a heating mantle attached, when you turn the device on, current goes through the small heating wire and the mantle. The mantle is now on *full blast* (120 V out of 120 V from the electric wall socket)! As the small heating wire warms the bimetal strip, the strip expands, distorts, and finally pulls the magnet from its metal plate, opening the circuit. The mantle now cools rapidly (0 V out of 120 V from the wall socket), along with the bimetal strip. Eventually, the strip cools enough to let the magnet get close to that metal plate, and— CLICK—everything's on full tilt again.

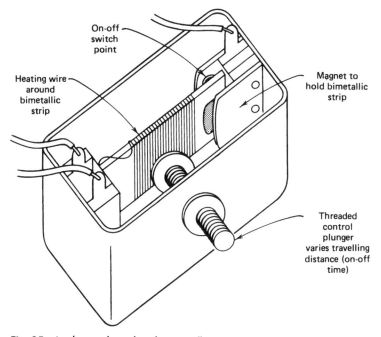

Fig. 85 Inside a mechanical stepless controller.

The front panel control varies the **duty cycle**, the time the controller is *full on*, to the time the controller is *full off*. If the flask, contents, and heating mantle are substantial, it takes a long time for them to warm up and cool down. A setup like that would have a large **thermal lag**. With small setups (approximately 50 ml or so), there is a small thermal lag and very wild temperature fluctuations can occur. Also, operating a heating mantle this way is just like repeatedly plugging and unplugging it directly into the wall socket. Not many devices easily take that kind of treatment.

3. *The electronic stepless controller.* Would you believe a *light dimmer*? The electronic controller has a *triac*, a semiconductor device, that lets fractions of the AC power cycle through to the heating mantle. The AC power varies like a sine wave, from 0 to 120 V from one peak to the next. At a setting of 25%, the triac remains off during much of the AC cycle, finally turning on when the time is right (Fig. 86). Although the triac does turn "full-off and full-on," it does so at times in so carefully controlled a way that the mantle never sees full line power (unless you deliberately set it there).

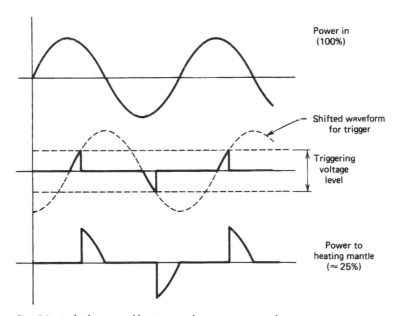

Fig. 86 Light dimmer and heating mantle triac power control.

CLAMPS
AND CLAMPING

Unfortunately, glass apparatus needs to be held in place with more than just spit and bailing wire. In fact, you would do well to use clamps. Life would be simple if there were just one type of fastener, but that's not the case.

1. ***The simple buret clamp*** (Fig. 87). Though popular in other chem labs, the simple buret clamp just doesn't cut it for organic lab. The clamp is too short, and adjusting angles with the "locknut" (by loosening the locknut, swiveling the clamp jaws to the correct angle, and tightening the locknut against the *back* stop, away from the jaws) is not a great deal of fun. If you're not careful, the jaws will slip right around and all the chemicals in your flask will fall out.

2. ***The simple extension clamp and clamp fastener*** (Fig. 88). This two-piece beast is the second best clamp going. It is much longer (approximately 12 inches), so you can easily get to complex setups. By loosening the **clamp holder thumbscrew**, the clamp can be pulled out, or pushed back, or rotated to any angle. By loosening the **ringstand thumbscrew**, the clamp, along with the clamp holder, can move up and down.

3. ***The three-fingered extension clamp*** (Fig. 89). This is truly the Cadillac of clamps with a price to match. They usually try to confuse

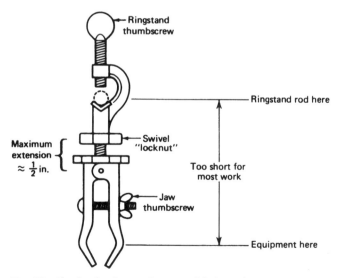

Fig. 87 The "barely adequate for organic lab" buret clamp.

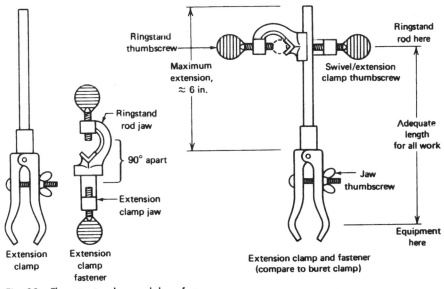

Ringstand thumbscrew

Maximum extension, ≈ 6 in.

Ringstand rod jaw

90° apart

Extension clamp jaw

Ringstand rod here

Swivel/extension clamp thumbscrew

Adequate length for all work

Jaw thumbscrew

Equipment here

Extension clamp

Extension clamp fastener

Extension clamp and fastener (compare to buret clamp)

Fig. 88 The extension clamp and clamp fastener.

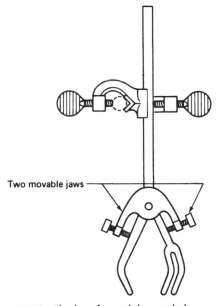

Two movable jaws

Fig. 89 The three-fingered clamp with clamp fastener.

19

you with *two thumbscrews* to tighten, unlike the regular extension clamp. This gives a bit more flexibility, at the cost of a slightly more complicated way of setting up. You can make life simple by opening the *two-prong bottom jaw to a 10° to 20° angle from the horizontal and treating that jaw as fixed*. This will save a lot of wear and tear when you set equipment up, but you can *always move the bottom jaw* if you have to.

CLAMPING A DISTILLATION SETUP

You'll have to clamp many things in your life as a chemist, and one of the more frustrating setups to clamp is the **simple distillation** (see Chapter 20, "Distillation"). If you can set this up, you probably will be able to clamp other common setups without much trouble. Here's one way to go about setting up the simple distillation.

1. OK, get a **ringstand** and an **extension clamp and a clamp fastener** and put them all together. What heat source? A Bunsen burner, and you'll need more room than you do with a heating mantle (see Chapter 18, "Sources of Heat"). In any case, you don't know where the receiving flask will show up, and then you might have to readjust the entire setup. Yes, you should have read the experiment before so you'd know about the heating mantles.

2. Clamp the flask (around the neck) a few inches up the ringstand (Fig. 90). We *are* using heating mantles, and you'll need the room underneath to drop the mantle in case it gets too hot. That's why the flask is *clamped at the neck. Yes. That's where the flask is ALWAYS clamped, no matter what heat source*, so it doesn't fall when the mantle comes down. What holds the mantle? Extension ring and clamp fastener.

3. Remember, whether you set these up from left to right, or right to left—*distilling flask first!*

4. Add the three-way adapter now (Fig. 91). Thermometer and thermometer adapter come later.

5. Put a second ringstand about one condenser length away from the first ringstand. Now add the condenser, just holding it onto the three-way adapter. Make a mental note of where the inner joint (end) of the condenser is—you'll want to put a clamp about there. Remove the condenser (Fig. 92).

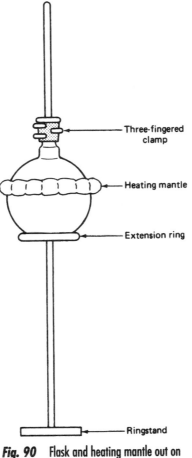

Three-fingered
clamp

Heating mantle

Extension ring

Ringstand

Fig. 90 Flask and heating mantle out on
a ringstand.

6. Get an extension clamp and clamp fastener. Open the jaws of the clamp so they're *wider* than the outer joint on a vacuum adapter. Set this clamp so that it can accept the outer joint of a vacuum adapter at the angle and height you made a mental note of in step 5.
7. Put a vacuum adapter on the end of your condenser and hold it there. Put the top of the condenser (outer joint) onto the three-way adapter, and get the vacuum adapter joint cradled in the clamp you've set for it. Push the condenser/vacuum adapter toward the three-way adapter; lightly tighten the clamp on the vacuum adapter. The clamp **must** stop the vacuum adapter from slipping *off the condenser* (Fig. 93).

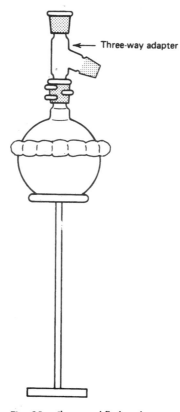

Fig. 91 Clamps and flask and three-way adapter.

8. At this point, you have two clamps, one holding the distilling flask and the other at the end of the condenser/vacuum adapter (Fig. 94).
9. There are two ways to go now:

    a. The receiving flask will not fit vertically from the end of the vacuum adapter to the benchtop (Fig. 94). Easy. With the vacuum adapter clamp **relaxed** (not so loose that the adapter falls off!), rotate the adapter toward you. There's no reason it *has* to be at a right angle. Stick a receiving flask on the end, put a suitable cork ring under the flask, and that's it. Note that you can **easily** change receiving flasks. Just slide one off and slide another on.

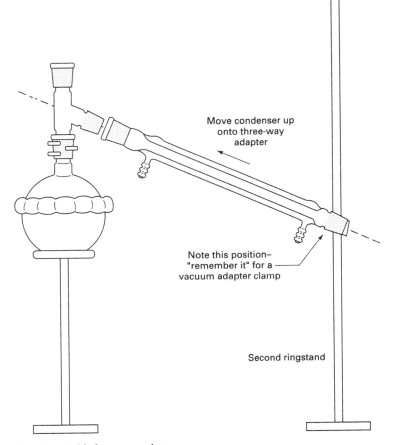

Move condenser up
onto three-way
adapter

Note this position–
"remember it" for a
vacuum adapter clamp

Second ringstand

Fig. 92 Trial fit for vacuum adapter.

19

b. There's a lot of room from the end of the vacuum adapter to the benchtop. In this case, you'll have to set up another clamp (possibly another ringstand) and clamp the flask at the neck (Fig. 95). This will make changing the receiving flask a bit more difficult; you'll have to unclamp one flask before slipping another in and re-clamping the new one.

10. All the clamps are set up, all the joints are tight—now where is that thermometer adapter?

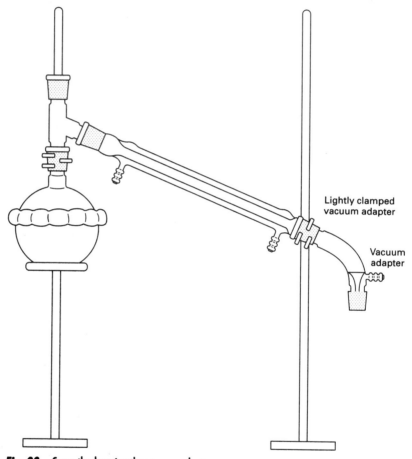

Fig. 93 Correctly clamping the vacuum adapter.

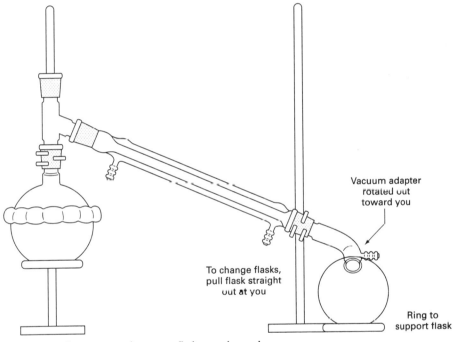

Vacuum adapter
rotated out
toward you

To change flasks,
pull flask straight
out at you

Ring to
support flask

Fig. 94 Distillation setup with receiving flask rotated toward you.

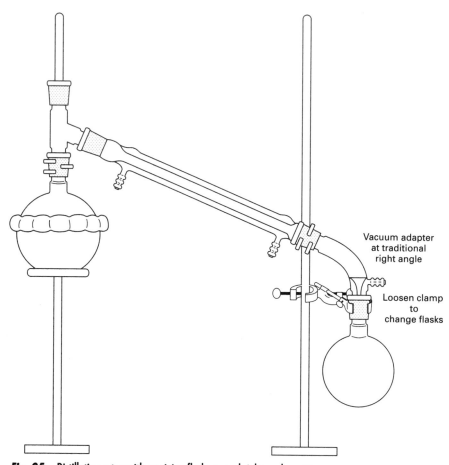

Vacuum adapter
at traditional
right angle

Loosen clamp
to
change flasks

Fig. 95 Distillation setup with receiving flask at usual right-angle position.

DISTILLATION

This separation or purification of liquids by vaporization and condensation is a very important step in one of our oldest professions. The word "still" lives on as a tribute to the importance of organic chemistry. There are two important points here.

1. **Vaporization.** Turning a liquid to a vapor.
2. **Condensation.** Turning a vapor to a liquid.

Remember these. They show up on quizzes.

But when do I use distillation? That is a very good question. Use the guidelines below to pick your special situation, and turn to that section. But you *should* read *all* the sections anyway.

1. **Class 1: Simple distillation.** Separating liquids boiling below 150°C at one atmosphere (1 atm) from

   a. Nonvolatile impurities.
   b. Another liquid boiling at least 25°C higher than the first. The liquids should dissolve in each other.

2. **Class 2: Vacuum distillation.** Separating liquids boiling above 150°C at 1 atm from

   a. Nonvolatile impurities.
   b. Another liquid boiling at least 25°C higher than the first. They should dissolve in one another.

3. **Class 3: Fractional distillation.** Separating liquid mixtures, soluble in each other, that boil at less than 25°C from each other at 1 atm.

4. **Class 4: Steam distillation.** Isolating tars, oils, and other liquid compounds that are *insoluble*, or slightly soluble, *in water at all temperatures*. Usually, natural products are steam distilled. They do *not* have to be liquids at room temperatures. (*E.g.,* caffeine, a solid, can be isolated from green tea.)

Remember, these are guides. If your compound boils at 150.0001°C, don't scream that you must do a vacuum distillation or both you and your product will die. I expect you to have some judgment and to pay attention to your instructor's specific directions.

DISTILLATION NOTES

1. Except for Class 4, steam distillation, two liquids that are to be separated must dissolve in each other. If they did not, they would form separable layers, which you could separate in a separatory funnel (see Chapter 15, "Extraction and Washing").
2. Impurities can be either **soluble** or **insoluble**. For example, the material that gives cheap wine its unique bouquet is soluble in the alcohol. If you distill cheap wine, you get clear grain alcohol separated from the "impurities," which are left behind in the distilling flask.

CLASS 1: SIMPLE DISTILLATION (FIG. 96)

For separation of liquids boiling below 150°C at 1 atm from

1. Nonvolatile impurities.
2. Another liquid boiling 25°C higher than the first liquid. *They must dissolve in each other.*

Sources of Heat

If one of the components boils below 70°C and you use a Bunsen burner, you may have a hard time putting out the fire. Use a steam bath or a heating mantle. Different distillations will require different handling (see Chapter 18, "Sources of Heat"). All the distillations always require heating, so Chapter 18 is really closely tied to this section. This goes for enlightenment on the use of **boiling stones** and **clamps** as well (see Chapter 17, "And Now—Boiling Stones" and Chapter 19, "Clamps and Clamping").

The Three-Way Adapter

If there is any *one place* your setup will fall apart, here it is (Fig. 97). When you set up the jointware, it is important that you have all the joints *line up*. This is tricky, since, as you push one joint together, another pops right out. If you're not sure, call your instructor. Let him inspect your work. Remember,

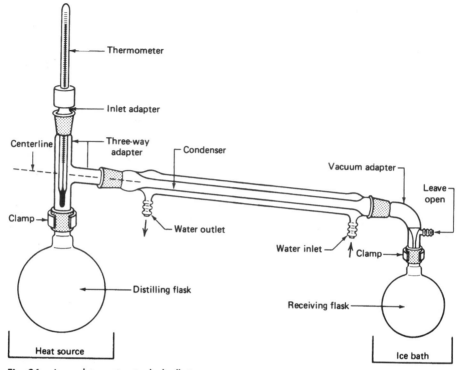

Fig. 96 A complete, entire simple distillation setup.

All joints must be tight!

The Distilling Flask

Choose a distilling flask carefully. If it's too big, you'll lose a lot of your product. If it's too small, you might have to distill in parts. *Don't fill the distilling flask more than half full.* Less than ⅓ full—you'll probably lose product. More than ½ full—you'll probably have undistilled material thrown up into the condenser (and into your previously clean product). Fill the distilling flask with the liquid you want to distill. You can remove the thermometer and thermometer adapter, fill the flask using a funnel, and then put the thermometer and its adapter back in place.

If you're doing a **fractional distillation** with a **column** (a Class 3 distillation), you should've filled the flask *before* clamping the setup. (Don't ever pour your mixture down a column. That'll contaminate everything!) You'll just have to disassemble some of the setup, fill the flask, reassemble

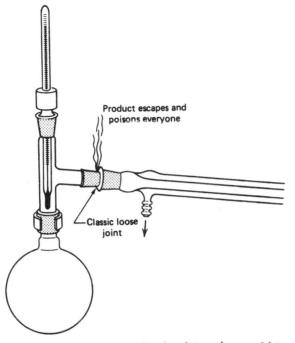

Product escapes and
poisons everyone

Classic loose
joint

Fig. 97 The commonly camouflaged until it's too late open joint.

what you've taken down, and pray that you haven't knocked all the other joints out of line.

Put in a boiling stone if you haven't already. These porous little rocks promote bubbling and keep the liquid from superheating and flying out of the flask. This flying around is called **bumping**. Never drop a boiling stone into hot liquid, or you may be rewarded by having your body soaked in the hot liquid as it foams out at you.

Make sure all the joints in your setup are tight. Start the heat s-l-o-w-l-y until gentle boiling begins and liquid starts to drop into the receiving flask at the rate of about 10 drops per minute. *This is important.* If nothing comes over, you're not distilling, but merely wasting time. You may have to turn up the heat to keep material coming over.

The Thermometer Adapter

Read all about it. Ways of having fun with thermometer adapters have been detailed (see text accompanying Figs. 22–26).

20

The Ubiquitous Clamp

A word about clamps. *Use!* They can save you $68.25 in busted glassware (see Chapter 19, "Clamps and Clamping").

The Thermometer

Make sure the entire thermometer bulb is *below the sidearm of the three-way adapter*. If you don't have liquid droplets condensing on the thermometer bulb, the temperature you read is *nonsense*. Keep a record of the temperature of the liquid or liquids that are distilling. It's a check on the purity. Liquid collected over a 2°C range is fairly pure. Note the similarity of this range with that of the *melting point* of a pure compound (see Chapter 12, "The Melting Point Experiment").

The Condenser

Always keep cold water running through the condenser, enough so that *at least the lower half is cold* to the touch. Remember that water should go *in the bottom* and *out of the top* (Fig. 96). Also, the water pressure in the lab may change from time to time and usually goes up at night, since little water is used then. So, if you are going to let condenser cooling water run overnight, tie the tubing on at the condenser and the water faucet with wire or something. And if you don't want to flood out the lab, see that the outlet hose can't flop out of the sink.

The Vacuum Adapter

It is important that the tubing connector remain *open to the air;* otherwise, the entire apparatus will, quite simply, explode.

Warning: Do not just stick the vacuum adapter on the end of the condenser and hope that it will not fall off and break.

This is foolish. I have no sympathy for anyone who will not use clamps to save their own breakage fee. They deserve it.

The Receiving Flask

The receiving flask should be large enough to collect what you want. You may need several, and they may have to be changed during the distillation. Standard practice is to have one flask ready for what you are going to throw away and others ready to save the stuff you want to save.

The Ice Bath

Why everyone insists on loading up a bucket with ice and trying to force a flask into this mess, I'll never know. How much cooling do you think you're going to get with just a few small areas of the flask barely touching ice? Get a suitable receptacle—a large beaker, enameled pan, or whatever. *It should not leak.* Put it under the flask. Put some water in it. *Now add ice.* Stir. Serves four.

Ice bath really means ice-water bath.

THE DISTILLATION EXAMPLE

Say you place 50 ml of liquid A (B.P. 50°C) and 50 ml of liquid B (B.P. 100°C) in a 250 ml R.B. flask. You drop in a boiling stone, fit the flask in a distillation setup, and turn on the heat. Bubbling starts, and soon droplets form on the thermometer bulb. The temperature shoots up *from room temperature to about 35°C,* and a liquid condenses and drips into the receiver. That's bad. The temperature should be close to 50°C. This low-boiling material is the **forerun** of a distillation, and you won't want to keep it.

Keep letting liquid come over until the temperature stabilizes at about 49°C. Quick! Change receiving flasks now!

The new receiving flask is on the condenser, and the temperature is about 49°C. Good. Liquid comes over, and you heat to get a rate of about 10 drops per minute collected in the receiver. As you distill, the temperature slowly increases to maybe 51°C and then starts moving up rapidly.

Here you stop the distillation and change the receiver. Now in one receiver you have a *pure liquid, B.P. 49-51°C.* Note this **boiling range**. It is just as good a test of purity as a melting point is for solids (see Chapter 12, "The Melting Point Experiment").

Always report a boiling point for liquids as routinely as you report melting points for solids. The boiling point is actually a **boiling range** and should be reported as such:

"B.P. 49–51°C"

If you now put on a new receiver, and start heating again, you may discover *more material coming over at 50°C*! Find that strange? Not so. All it means is that you were distilling too rapidly and some of the low-boiling material was left behind. It is very difficult to avoid this situation. Sometimes it is best to ignore it, unless a yield is very important. You can combine this "new" 50°C fraction with the other good fraction.

For liquid B, boiling at 100°C, merely substitute some different boiling points and go over the same story.

THE DISTILLATION MISTAKE

OK, you set all this stuff up to do a distillation. Everything's going fine. Clamps in the right place. No arthritic joints, even the vacuum adapter is clamped on, and the thermometer is at the right height. There's a bright golden haze on the meadow, and everything's going your way. So, you begin to boil the liquid. You even remembered the boiling stone. Boiling starts slowly, then more rapidly. You think, "This is *it*!" Read that temperature, now. Into the notebook: "The mixture started boiling at 26°C."

And you are dead wrong.

What happened? Just ask—

Is there liquid condensing on the thermometer bulb??
No!

So, congratulations, you've just recorded the room temperature. There are days when over half of the class will report distillation temperatures as "Hey I see it start boiling now" temperatures. Don't participate. Just keep watching as the liquid boils. Soon, droplets *will* condense on the thermometer bulb. The temperature will go up quickly, and then *stabilize*. Now read the temperature. That's the boiling point. But wait! It's *not* a distillation temperature until that first drop of liquid falls into the receiving flask.

CLASS 2: VACUUM DISTILLATION

For separation of liquids boiling above 150°C at 1 atm from

1. Nonvolatile impurities.
2. Another liquid boiling 25°C higher than the first liquid. They must dissolve in each other. This is like the **simple distillation** with the changes shown (Fig. 98).

Why vacuum distill? If the substances boil at high temperatures at 1 atm, they may decompose when heated. Putting a vacuum over the liquid makes the liquid boil at a lower temperature. With the pressure reduced, there are fewer molecules in the way of the liquid you are distilling. Since the molecules require less energy to leave the surface of the liquid, *you can distill at a lower temperature*, and *your compound doesn't decompose.*

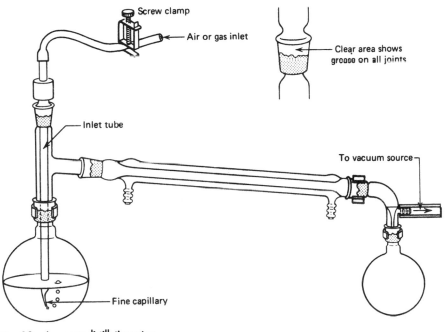

Fig. 98 A vacuum distillation setup.

Pressure Measurement

If you want to measure the pressure in your vacuum distillation setup, you'll need a **closed-end manometer**. There are a few different types, but they all work essentially the same way. I've chosen a "stick" type (Fig. 99). This particular model needs help from a short length of rubber tubing and a glass T to get connected to the vacuum distillation setup.

1. Turn on the source of vacuum and wait a bit for the system to stabilize.
2. Turn the knob on the manometer so that the notch in the joint lines up with the inlet.
3. Wait for the mercury in the manometer to stop falling.
4. Read the *difference* between the inner and outer levels of mercury. This is the system pressure, literally in **millimeters of mercury**, which we now call *torr*.
5. Turn the knob on the manometer to disconnect it from the inlet. Don't leave the manometer permanently connected. Vapors from your distillation, water vapor from the aspirator, and so on, may contaminate the mercury.

Manometer Hints

1. Mercury is toxic, the vapor from mercury is toxic, mercury spilled breaks into tiny globules that evaporate easily and are toxic, it'll alloy with your jewelry, and so on. Be very careful not to expose yourself (or anyone else) to mercury.
2. If the mercury level in the *inner* tubes goes *lower than that of the outer tube,* it does not mean that you have a negative vacuum. Some air or other vapor has gotten into the inner stick, and with the vacuum applied, the vapor expands and drives the mercury in the inner tube lower than that in the outer tube. This manometer is unreliable, and you should seek a replacement.
3. If a rubber tube connected to the vacuum source and the system (or manometer) *collapses,* you've had it. The system is no longer connected to the vacuum source, and as air from the bleed tube or vapor from the liquid you're distilling fills your distillation setup, the pressure in the system goes up. Occasionally, test the vacuum hoses, and if they collapse under vacuum, replace them with sturdier hoses that can take it.

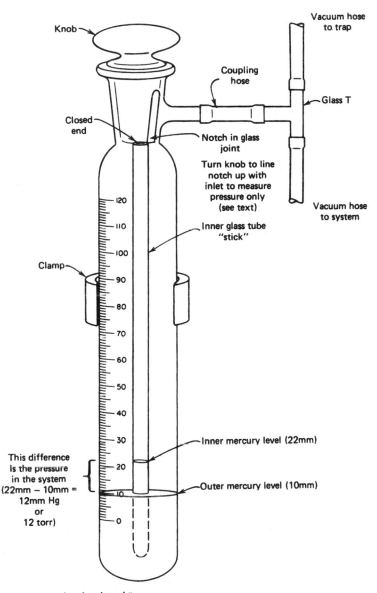

Knob

Coupling
hose

Vacuum hose
to trap

Glass T

Closed
end

Notch in glass
joint

Turn knob to line
notch up with
inlet to measure
pressure only
(see text)

Vacuum hose
to system

Inner glass tube
"stick"

Clamp

120

110

100

90

80

70

60

50

40

30

Inner mercury level (22mm)

This difference
Is the pressure
in the system
(22mm − 10mm =
12mm Hg
or
12 torr)

20

Outer mercury level (10mm)

10

0

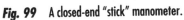

Fig. 99 A closed-end "stick" manometer.

20

Leaks

Suppose, by luck of the draw, you've had to prepare and purify 1-octanol (B.P. 195°C). You know that, if you simply distill 1-octanol, you run the risk of having it decompose, so you set up a vacuum distillation. You hook your setup to a water aspirator and water trap, and you attach a closed-end "stick" manometer. You turn the water for the aspirator on full blast and open the stick manometer. After a few minutes, nothing seems to be happening. You *pinch* the tubing going to the vacuum distillation setup (but not to the manometer), closing the setup off from the source of vacuum. Suddenly, the mercury in the manometer starts to drop. You *release* the tube going to the vacuum distillation setup, and the mercury jumps to the upper limit. You have **air leaks** in your vacuum distillation setup.

Air leaks can be difficult to find. At best, you push some of the joints together again and the system seals itself. At worst, you have to take apart all the joints and regrease every one. Sometimes you've forgotten to grease all the joints. Often a joint has been etched to the point that it cannot seal under vacuum, when it is perfectly fine for other applications. Please get help from your instructor.

Pressure and Temperature Corrections

You've found all the leaks, and the pressure in your vacuum distillation setup is, say, 25 torr. Now you need to know the boiling point of your compound, 1-octanol, this time at 25 torr, and not 760 torr. You realize it'll boil at a lower temperature, but just how low? The handy nomographs can help you estimate the new boiling point.

This time you have the boiling point at 760 torr (195°C) and the pressure you are working at (25 torr) so using Fig. 100a, you

1. Find the boiling point at 760 (195°C) on line B (the middle one).
2. Find the pressure you'll be working at (25 torr) on line C (the one on the far right). You'll have to estimate this point.
3. Using a straightedge, line up these two points and see where the straightedge cuts the observed boiling point line (line A, far left). I get about 95°C.

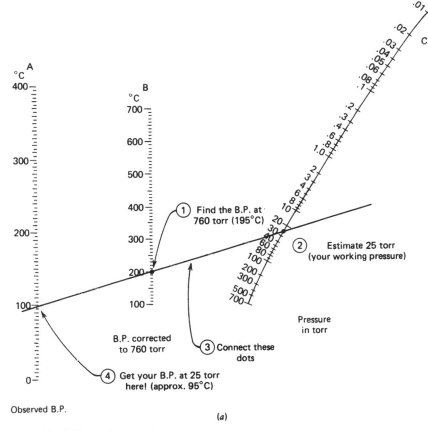

Fig. 100 (a) One point conversion.

So a liquid that boils at 195°C at 760 torr will boil at about 95°C at 25 torr. Remember, this is an estimate.

Now suppose you looked up the boiling point of 1-octanol and all you found was: 98^{19}. This means that the boiling point of 1-octanol is 98°C at 19 torr. Two things should strike you.

1. This is a *higher* boiling point at a *lower* pressure than we've gotten from the nomograph.
2. I wasn't kidding about this process being an estimation of the boiling point.

20

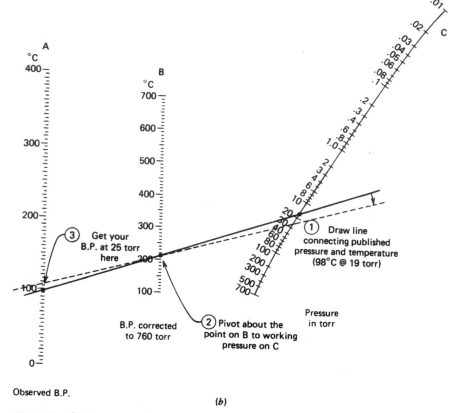

Fig. 100 (b) Two point conversion.

Now we have a case of having an observed boiling point at a pressure that is *not* 760 torr (1-octanol again; 98°C at 19 torr). We'd like to get to 25 torr, our working pressure. This requires a double conversion, as shown in Figure 100b.

1. On the observed B.P. line (line A) find 98°C.
2. On the pressure in torr line (line C) find 19.
3. Using a straightedge, connect those points. Now read the B.P. corrected to 760 (line B): I get 210°C.
4. Now, using the 210°C point as a fulcrum, pivot the straightedge until the 210°C point on line B and the pressure you're working at (25 torr) on line C line up. You see, you're in the same position as in the previous example with a "corrected to 760 torr B.P." and a working pressure.
5. See where the straightedge cuts the observed boiling point line (line A). I get 105°C.

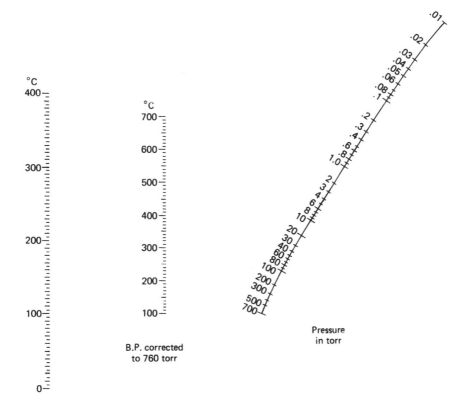

°C
400—
300—
200—
100—
0—

Observed B.P.

°C
700—
600—
500—
400—
300—
200—
100—

B.P. corrected
to 760 torr

Pressure
in torr

Fig. 101 A clean nomograph for your own use.

So, we've estimated the boiling point to be about 105°C at 25 torr. The last time it was 95°C at 25 torr. Which is it? Better you should say you *expect* your compound to come over from 95–105°C. Again, this is not an unreasonable expectation for a vacuum distillation.

The pressure-temperature nomograph is really just a simple, graphical application of the Clausius-Clapeyron equation. If you know the heat of vaporization of a substance, as well as its normal boiling point, you can calculate the boiling point at another temperature. You do have to assume that the heat of vaporization is constant over the temperature range you're working with, and that's not always so. Where's the heat of vaporization in the nomograph? One is built into the slopes and spacings on the paper. And, yes, that means that the heat of vaporization is forced to be the same for all compounds, be they alkanes, aldchydes, or ethers. So do not be surprised at the inaccuracies in this nomograph; be amazed that it works as well as it does.

Vacuum Distillation Notes

1. Read all the notes on class 1.

2. The thermometer can be replaced by a **gas inlet tube**. It has a long, fine capillary at one end (Fig. 98). This is to help stop the *extremely bad bumping* that goes along with vacuum distillations. The fine stream of bubbles through the liquid produces the same results as a boiling stone. Boiling stones are useless, since all the adsorbed air is whisked away by the vacuum and the nucleating cavities plug up with liquid. The fine capillary does not let in a lot of air, so we are doing a vacuum distillation anyway. Would you be happier if I called it a **reduced-pressure** distillation? An inert gas (nitrogen?) may be let in if the compounds decompose in air.

3. If you can get a **magnetic stirrer** and **magnetic stirring bar,** you won't have to use the gas inlet tube approach. Put a magnetic stirring bar in the flask with the material you want to vacuum distill. Use a *heating mantle* to heat the flask, and put the magnetic stirrer under the mantle. When you turn the stirrer on, a magnet in the stirrer spins, and the stirring bar (a Teflon-coated magnet) spins. Admittedly, stirring through a heating mantle is not easy, but it can be done. *Stirring the liquid also stops the bumping.*

Remember, first the stirring, then the vacuum, then the heat— or WOOSH! Got it?

4. Control of heating is *extremely critical.* I don't know how to shout this loudly enough on paper. Always apply the vacuum first and watch the setup for awhile. Air dissolved or trapped in your sample or a highly volatile leftover (maybe ethyl ether from a previous extraction) can come flying out of the flask *without the heat.* If you heated such a setup a bit and then applied the vacuum, your sample would blow all over, possibly right into the receiving flask. Wait for the contents of the distilling flask to calm down before you start the distillation.

5. If you know you have low-boiling material in your compound, think about distilling it at atmospheric pressure first. If, say, half the liquid you want to vacuum distill is ethyl ether from an extraction, consider doing a simple distillation to get rid of the ether. Then the ether (or any other low-boiling compound) won't be around to cause trouble during the vacuum distillation. If you distill first at 1 atm, *let the flask cool before you apply the vacuum.* Otherwise, your compound will fly all

over and probably will wind up, undistilled and impure, in your receiving flask.

6. *Grease all joints, no matter what* (see "Greasing the Joints" in Chapter 4). Under vacuum, it is easy for any material to work its way into the joints and turn into concrete, and the joints will never, ever come apart again.

7. The **vacuum adapter** is connected to a **vacuum source**, either a **vacuum pump** or a **water aspirator**. Real live vacuum pumps are expensive and rare and not usually found in the undergraduate organic laboratory. If you can get to use one, that's excellent. See your instructor for the details. The **water aspirator** is used lots, so read up on it.

8. During a vacuum distillation, it is not unusual to collect a *pure compound over a 10–20°C temperature range*. If you don't believe it, you haven't ever done a vacuum distillation. It has to do with pressure changes throughout the distillation because the setup is far from perfect. Although a vacuum distillation is not difficult, it requires peace of mind, large quantities of patience, and a soundproof room to scream in so you won't to disturb others.

9. A **Claisen adapter** in the distilling flask lets you take temperature readings and can help stop your compound from splashing over into the distillation receiver (Fig. 102). On a microscale, though, the extra length of the Claisen adapter could hold up a lot of your product. Check with your instructor. Also, you could use a **three-neck flask** (Fig. 103). Think! And, of course, use some glassware too.

CLASS 3: FRACTIONAL DISTILLATION

For separation of liquids, soluble in each other, that boil less than 25°C from each other, use fractional distillation. This is like simple distillation with the changes shown (Fig. 80).

    Fractional distillation is used when the components to be separated boil *within 25° of each other*. Each component is called a **fraction**. Clever where they get the name, eh? This temperature difference is not gospel. And don't expect terrific separations either. Let's just leave it at *close boiling points*. How close? That's hard to answer. Is an orange? That's easier to answer. If the experiment tells you to "fractionally distill," at least you'll be able to set it up right.

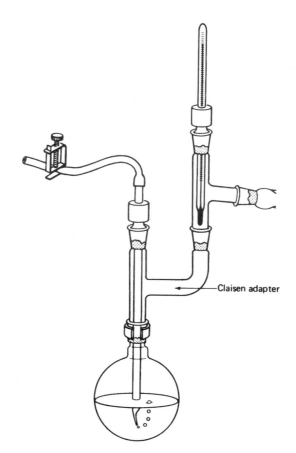

Fig. 102 A Claisen adapter so you can vacuum distill and take temperatures too.

How This Works

If one distillation is good, two is better. And fifty, better still. So you have lots and lots of little, tiny distillations occurring on the surfaces of the **column packing**, which can be glass beads, glass helices, ceramic pieces, metal chips, or even stainless-steel wool.

As you heat your mixture it boils, and the vapor that comes off this liquid is *richer in the lower-boiling component*. The vapor moves out of the flask and condenses, say, on the first centimeter of column packing. Now, the composition of the liquid still in the flask has changed a bit—it is *richer in*

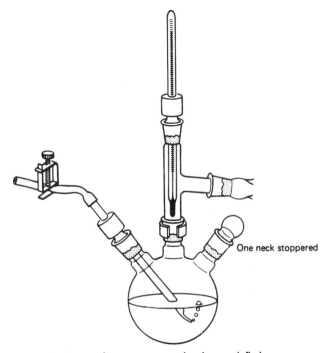

One neck stoppered

Fig. 103 Some multipurpose setup with a three-neck flask.

the higher-boiling component. As more of *this* liquid boils, more hot vapor comes up, mixes with the first fraction, and produces a new vapor of different composition—*richer yet in the more volatile (lower boiling) component.* And guess what? This new vapor condenses in the *second centimeter* of column packing. And again, and again, and again.

Now all these are **equilibrium steps**. It takes some time for the fractions to move up the column, get comfortable with their surroundings, meet the neighbors....And if you *never* let any of the liquid-vapor mixture out of the column, a condition called **total reflux**, you might get a single pure component at the top—namely, the lower boiling, more volatile component all by itself! This is an ideal separation.

Fat lot of good that does you when you have to hand in a sample. So, you turn up the heat, let some of the vapor condense, and *take off this top fraction.* This raises hell in the column. *Nonequilibrium conditions abound— mixing. Arrrgh!* No more completely pure compound. And the faster you distill, the faster you let material come over, the higher your **through-**

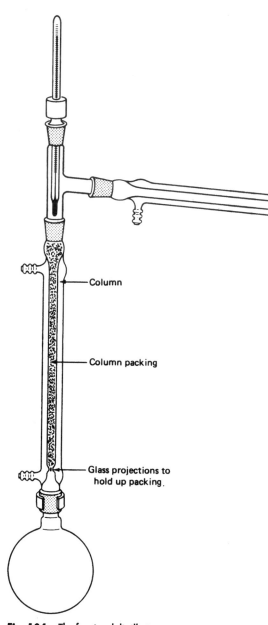

Column

Column packing

Glass projections to hold up packing.

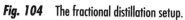

Fig. 104 The fractional distillation setup.

put—the worse this gets. Soon you're at **total takeoff,** and there is no time for an equilibrium to get established. And if you're doing that, you shouldn't even bother using a column.

You must strike a compromise. Fractionally distill as slowly as you can, keeping in mind that eventually the lab does end. Slow down your fractional distillations; I've found that 5–10 drops per minute coming over into the receiving flask is usually suggested. It will take a bit of practice before you can judge the best rate for the best separation. See your instructor for advice.

Fractional Distillation Notes

1. Read all the notes on Class 1.
2. Make *sure* you have not confused the **column** with the **condenser**. The *column is wider and has glass projections inside*, at the bottom, to hold up the packing.
3. *Don't break off the projections!*
4. Do not run water through the jacket of the column!
5. Sometimes, the column is used *without* the column packing. This is all right, too.
6. If it is necessary, and it usually is, push a wad of heavy metal wood down the column, *close to the support projections*, to support the packing chips. Sometimes the packing is entirely this stainless-steel wool. You can see that it is self-supporting.
7. Add the column packing. Shake the column lightly to make sure none of the packing will fall out later into your distillation.
8. With all the surface area of the packing, a lot of liquid is *held up* on it. This phenomenon is called **column holdup**, *since it refers to the material retained in the column. Make sure you have enough compound to start with, or it will all be lost on the packing.*
9. A **chaser solvent** or **pusher solvent** is sometimes used to help blast your compound off the surface of the packing material. It should have a *tremendously high boiling point relative to what you were fractionating*. After you've collected most of one fraction, some of this material is left on the column. So, you throw this chaser solvent into the distillation flask, fire it up, and start to distill the chaser solvent. As the chaser solvent comes up the column, it heats the packing material, your compound is blasted off the column packing, and more of your compound comes over. Stop collecting when the temperature starts to

rise—that's the chaser solvent coming over now. As an example, you might expect p-xylene (B.P. 138.4°C) to be a really good chaser, or pusher, for compounds that boil less than, say, 100°C.

But you have to watch out for the deadly azeotropes.

AZEOTROPES

Once in a while, you throw together two liquids and find that you cannot separate part of them. And I don't mean because of poor equipment, or poor technique, or other poor excuses. You may have an **azeotrope**, a mixture with a *constant boiling point*.

One of the best known examples is the ethyl alcohol-water azeotrope. This 96% alcohol-4% water solution will boil to dryness, at a *constant temperature*. It's slightly scary, since you learn that a liquid is a pure compound if it boils at a constant temperature. And you thought you had it made.

There are two types of azeotrope. If the azeotrope boils off first, it's a **minimum-boiling azeotrope**. After it's all gone, if there is any other component left, only then will that component distill.

If any of the components come off first, and then the azeotrope, you have a **maximum-boiling azeotrope**.

Quiz question:

Fifty milliliters of a liquid boils at 74.8°C from the beginning of the distillation to the end. Since there is no wide boiling range, can we assume that the liquid is pure?

No. It may be a constant boiling mixture called an azeotrope.

You should be able to see that you have to be really careful in selecting those chaser or pusher solvents mentioned. Sure, water (B.P. 100°C) is hot enough to chase ethyl alcohol (B.P. 78.3°C) from any column packing. Unfortunately, water and ethyl alcohol form an azeotrope and the technique won't work. (*Please* see Chapter 34, "Theory of Distillation.")

CLASS 4: STEAM DISTILLATION

Mixtures of tars and oils must not dissolve well in water (well, not much, anyway), so we can steam distill them. The process is pretty close to simple distillation, but you should have a way of getting *fresh hot water into the setup* without stopping the distillation.

Why steam distill? If the stuff you're going to distill is *only slightly soluble in water* and may decompose at its boiling point and the bumping will be terrible with a vacuum distillation, it is better to **steam distill**. Heating the compound in the presence of steam makes the compound boil at a lower temperature. This has to do with partial pressures of water and organic oils and such.

There are two ways of generating steam: externally and internally.

External Steam Distillation

In an **external steam distillation**, you lead steam from a steam line, through a water trap, and thus into the system. The steam usually comes from a **steam tap** on the benchtop. This is classic. This is complicated. This is dangerous.

1. Set up your external steam distillation apparatus in its entirety. Have *everything* ready to go. Have the substance you want to distill already in the distilling flask. This includes having the material you want to distill in the distilling flask, the steam trap already attached, condensers up and ready, a large receiving flask, and so on. All you should have to do is attach a single hose from the steam tap to your steam trap and start the steam.

2. *Have your instructor check your setup before you start!* I cannot shout this loudly enough on this sheet of paper. Interrupting an external steam distillation, just because you forgot your head this morning, is a real trial.

3. Connect a length of rubber tubing to your bench steam outlet and lead the rubber tubing into a drain.

4. Now, watch out! Slowly, carefully, open the steam stopcock. Often you'll hear clanging, bonking, and thumping, and a mixture of rust, oil, and dirt-laden water will come spitting out. Then some steam bursts come out, and finally, you have a stream of steam. Congratulations. You have just *bled the steam line*. Now close the steam stopcock, wait for the rubber tubing to cool a bit, and then...

5. Carefully (**Caution**—may be hot!) attach the rubber tubing from the steam stopcock to the inlet of your steam trap.

6. Open the steam trap drain, then carefully reopen the bench steam stopcock. Let any water drain out of the trap, then carefully close the drain clamp. Be careful.

7. You now have steam going through your distillation setup, and as soon as product starts to come over, you'll be doing an external steam distillation. Periodically, open the steam trap drain (**Caution**—hot!) and let the condensed steam out.

8. Apparently, you can distill as fast as you can let the steam into your setup, *as long as all the steam condenses* and doesn't go out into the room. Sometimes you need to hook two condensers together, making a very long supercondenser, when you steam distill. Check with your instructor.

9. When you're finished (see "Steam Distillation Notes" following), turn off the steam, let the apparatus cool, and dismantle everything.

You can use many types of steam traps with your distillation setup. I've shown two (Fig. 105a, b), but these are not the only ones, and you may use something different. The point is to note the **steam inlet** and the **trap drain**, and how to use them.

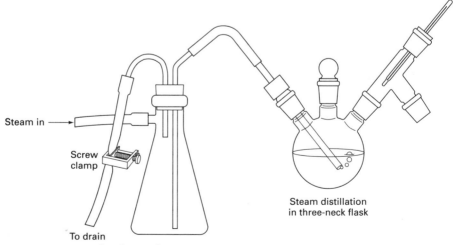

Steam in

Screw clamp

To drain

Steam distillation in three-neck flask

Fig. 105 (a) Examples of external steam traps.

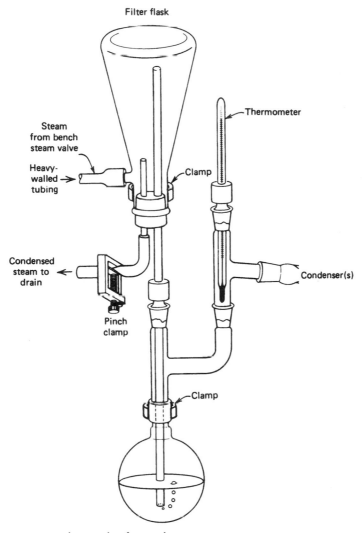

Fig. 105 (b) Examples of external steam traps.

20

Internal Steam Distillation

1. You can add hot water to the flask (Fig. 106) that will generate steam and thus provide an **internal source of steam**. This method is used almost exclusively in an undergraduate organic lab for the simple reason that it is so simple.
2. Add to the distilling flask at least three times as much water (maybe more) as sample. Do not fill the flask much more than half full (three quarters, maybe). You've got to be careful. Very careful.
3. Periodically, add more *hot water* as needed. When the water boils and turns to steam, it also leaves the flask, carrying product.

Steam Distillation Notes

1. Read all the notes on Class 1 distillations.
2. Collect some of the distillate, the stuff that comes over, in a small test tube. Examine the sample. If you see *two layers*, or *if the solution is cloudy*, you're not done. Your product is still coming over. Keep distilling and keep adding hot water to generate more steam. *If you*

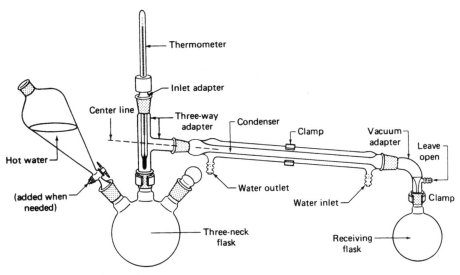

Fig. 106 An internal steam distillation setup.

don't see any layers, don't assume you're done. If the sample is slightly soluble in the water, the two layers or cloudiness might not show up. Try *salting-out*. This has been mentioned before in connection with **extraction** and **recrystallization** as well (see "Salting-Out" in Chapter 13 and "Extraction Hints" in Chapter 15). Add some salt to the solution you've collected in the test tube, shake the tube to dissolve the salt, and, if you're lucky, more of your product may be squeezed out of the aqueous layer, forming a *separate layer*. If that happens, *keep steam distilling* until the product does not come out when you treat a test solution with salt.

3. There should be two layers of liquid in the receiving flask at the end of the distillation. One is *mostly water*; the other is *mostly product*. To find out which is which, add a small quantity of water to the flask. The water will go into the water layer. (Makes sense.) Be very careful with this test, however; it is sometimes very hard to tell where the water has gone.

4. If you have to get more of your organic layer out of the water, you can do a **back-extraction** with an immiscible solvent (see "The Road to Recovery—Back-Extraction" in Chapter 15).

20

MICROSCALE DISTILLATION

LIKE THE BIG GUY

Before you read this chapter, read up on Class 1: Simple Distillation (Chapter 20) so you'll know what I'm talking about.

Until recently, there was no way you could duplicate an ordinary distillation of any class in microscale (re-read the material on the different classes in Chapter 20, "Distillation"), because the equipment either didn't exist or was fabricated with glass joints that didn't match up. Recently, Ace Glass (and others for all I know) has produced an all ℑ 14/10 joint microscale kit. Unfortunately neither a three-way nor a vacuum takeoff adapter is included. These pieces are, however, available separately.

So you *can* set up the same apparatus and do the same distillation for the same classes you've always done.

That's not completely true. Using this apparatus at the true microscale (0.01 mole), you have two drops of a liquid to distill. I'd see my instructor before I did any of these techniques with fewer than 10 ml.

Class 1: Simple Distillation

Perhaps you'll be using a conical vial rather than an R.B. flask, but the setup and operation are the same as with the larger apparatus. **Note:** the tubing carrying cooling water can be so unwieldy that moving the tubing can pull this tiny setup out of alignment.

Class 2: Vacuum Distillation

Because the setup is so small and your sample size is so small, you've got to be very careful here. With ℑ 14/20 ware (it's a little bigger), the operations are identical to those of the ℑ 24/40 or ℑ 19/22 sizes. I don't *have* the ℑ 14/10 setup yet, so I can only extrapolate. You'll have the same entertaining experiences. Watch the vacuum hose, too. It can be so unwieldy that it pulls the apparatus out of line if it's moved.

Oh. Check out the "O-ring cap seal" discussion for a perspective on microscale vacuum distillation (see Chapter 6, "Microscale Jointware").

Class 3: Fractional Distillation

You'll probably substitute one of those **air condensers** for the distilling column. As a straight, unjacketed tube, the air condenser is not very efficient. I'd be careful about packing it with anything; the column holdup, as small as it is, is likely to hold up a lot of your product.

Class 4: Steam Distillation

With sample sizes this small, internal steam distillation is the only way out. With the microscale equipment, you can inject water by syringe rather than opening the apparatus up to pour water in, and lose less product. Again, if your sample is very small, see your instructor about any problems.

MICROSCALE DISTILLATION II: THE HICKMAN STILL

If you have only 1–2 ml to distill, then the Hickman still is probably your only way out (Fig. 107).

The Hickman Still Setup

1. Set up the stirring hot plate and sand bath. I suppose you *could* stick a thermometer in the sand to give you an idea of how warm the sand is.
2. Put the sample you want to distill into an appropriately sized conical vial. Two things: Don't fill the vial more than halfway, and don't use such a big vial that the product disappears. Less than one third full, and you'll lose a lot on the walls of the vial.
3. These vials can fall over easily. Keep them in a holder, or clamp them until you're ready to use them.
4. Add a microboiling stone *or* a spin vane, not both, to the vial.
5. Get a clamp ready to accept the Hickman still at about the position you want; just open and loosely clamped—horizontally—to the ringstand.
6. Scrunch the conical vial down into the sand, and push it to where you want it to be. Hold the vial at the top.

7. Put the Hickman still on the vial. Now, even though you have a sand buttress, this baby is still top heavy and can fall. Hold the vial **and** the still somewhere about where they meet.
8. Jockey the clamp and/or the vial and still so that they meet. Clamp the still at the top in the clamp jaws. Get the vial and still in their final resting place; tighten the clamp to the ringstand.
9. Hang a thermometer by a clamp and stopper so that the thermometer goes straight down the throat of the still. Place the thermometer bulb at about the joint of the vial and still. Do *not* let the thermometer touch the sides of the vial or still.
10. You're ready to heat.

Hickman Still Heating

You don't have very much liquid here, so it doesn't take much heat to boil your sample. If you have no idea at what temperature your sample will boil at, you should just heat carefully until you see it boil, and then *immediately* turn the heat down 10–20%. Of course, you have a microboiling stone or spinning spin vane in there, don't you?

As you heat the sample, you'll see a ring of condensate travel up the vial, and when it hits the thermometer bulb, the temperature reading should move up. As usual, if there's **refluxing liquid** on the bulb and the temperature has stabilized, that's the boiling point.

The vapor then travels up into the still head, condenses on the glass surface, and falls down, to be caught in the bulge of the still (Fig. 107).

When you feel you've collected enough liquid (this will depend on the experiment), you can go after it with a pipet and rubber bulb.

Recovering Your Product

If you have a thermometer down the throat of the **Hickman still,** forget about getting your product out with a disposable pipet; there's just not enough room, period. If you use a 5-¾ inch pipet, you'll likely snap the tip off in the still when you try to bend it into reaching for your product, caught in the bulge of the still. Yes, you *can* just about wangle the end of a 9 inch pipet into the crevice and pick up your product, but there's a good chance

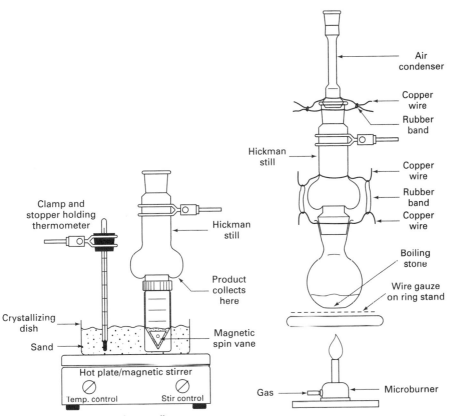

Fig. 107 Two different Hickman Still setups.

of breaking the tip off anyway. You might wind up having to put a bend in the tip, using a burner and your considerable glassworking skills. You have to use a small burner, and a small flame, and heat the tube carefully so it doesn't close off or melt off—in short, see your instructor about this. You should wind up with a slightly bent tip at the end of the 9 inch pipet. What I haven't yet tried, and wonder why it hasn't been done yet, is to put a small length of the gas-collecting tubing—very thin, narrow plastic or Teflon tubing—on the end of a pipet bent in a curve so you can go after your distillation product without resorting to glassworking. Again, the 9 inch pipet *can* just make it, but be very careful.

REFLUX

Just about 80% of the reactions in organic lab involve a step called **refluxing.** You use a reaction solvent to keep materials dissolved and at a constant temperature by boiling the solvent, condensing it, and returning it to the flask.

For example, say you have to heat a reaction to around 80°C for 17 hours. Well, you can stand there on your flat feet and watch the reaction all day. Me? I'm off to the **reflux**.

Usually, you'll be told what solvent to use, so selecting one should not be a problem. What happens more often is that you choose the reagents for your particular synthesis, put them into a solvent, and **reflux** the mixture. You boil the solvent and condense the solvent vapor *so that ALL the solvent runs back into the reaction flask* (see "Class 3: Fractional Distillation" in Chapter 20). The *reflux temperature is near the boiling point of the solvent.* To execute a reflux,

1. Place the reagents in a round-bottomed flask. The flask should be large enough to hold both the reagents and enough solvent to dissolve them, without being much more than half full.

2. You should now choose a solvent that

   a. Dissolves the reactants at the boiling temperature.
   b. Does *not* react with the reagents.
   c. Boils at a temperature that is high enough to cause the desired reaction to go at a rapid pace.

3. Dissolve the reactants in the solvent. Sometimes *the solvent itself is a reactant.* Then don't worry.

4. Place a condenser, upright, on the flask, connect the condenser to the water faucet, and run water through the condenser (Fig. 108). Remember—in at the bottom and out at the top.

5. Put a suitable heat source under the flask and adjust the heat so that the solvent condenses *no higher than halfway up the condenser.* You'll have to stick around and watch for a while, since this may take some time to get started. Once the reaction is stable, though, go do something else. You'll be ahead of the game for the rest of the lab.

6. Once this is going well, leave it alone until the reaction time is up. If it's an overnight reflux, wire the water hoses on so they don't blow off when you're not there.

7. When the reaction time is up, turn off the heat, let the setup cool, dismantle it, and collect and purify the product.

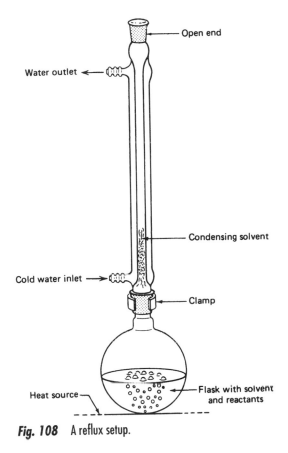

Fig. 108 A reflux setup.

A DRY REFLUX

If you have to keep the atmospheric water vapor out of your reaction, you must use a **drying tube** and **the inlet adapter** in the reflux setup (Fig. 109). You can use these if you need to keep water vapor out of any system, not just the reflux setup.

1. If necessary, *clean and dry the drying tube*. You don't have to do a thorough cleaning unless you suspect that the anhydrous drying agent is *no longer anhydrous*. If the stuff is caked inside the tube, it is probably dead. You should clean and recharge the tube at the beginning of the semester. Be sure to use *anhydrous* calcium chloride or sulfate. It should last one semester. If you are fortunate, **indicating Drierite**, a specially prepared anhydrous calcium sulfate, might

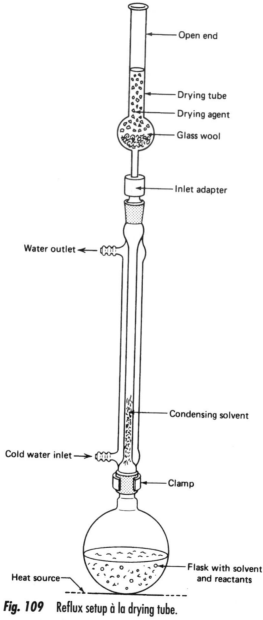

Open end

Drying tube

Drying agent

Glass wool

Inlet adapter

Water outlet

Condensing solvent

Cold water inlet

Clamp

Flask with solvent
and reactants

Heat source

Fig. 109 Reflux setup à la drying tube.

22

be mixed in with the white Drierite. If the color is *blue*, the drying agent is good; if *red*, the drying agent is no longer dry, and you should get rid of it (see Chapter 10, "Drying Agents").

2. Put in a loose plug of *cotton* to keep the drying agent from falling into the reaction flask.

3. Assemble the apparatus as shown, with the *drying tube and adapter on top of the condenser*.

4. At this point, reagents may be added to the flask and heated with the apparatus. Usually, the apparatus is heated while empty to drive water off the walls of the apparatus.

5. Heat the apparatus, usually empty, on a steam bath, giving the entire setup a quarter-turn every so often to heat it evenly. A burner can be used *if there is no danger of fire* and if heating is done carefully. The heavy ground-glass joints will crack if they are heated too much.

6. Let the apparatus cool to room temperature. As it cools, air is drawn through the drying tube before it hits the apparatus. The moisture in the air is trapped by the drying agent.

7. Quickly add the dry reagents or solvents to the reaction flask, and reassemble the system.

8. Carry out the reaction as usual like a standard reflux.

ADDITION AND REFLUX

Every so often you have to add a compound to a setup while the reaction is going on, usually along with a **reflux**. Well, in adding new reagents you *don't break open the system, let toxic fumes out, and make yourself sick*. You use an **addition funnel**. Now, we talked about addition funnels back with **separatory funnels** (Chapter 15) when we were considering the **stem,** and that might have been confusing.

Funnel Fun

Look at Fig. 110a. It is a true sep funnel. You put liquids in here and shake and extract them. But could you use this funnel to add material to a setup? *NO*. No ground glass joint on the end, and only glass joints fit glass joints. Right? Of course, right.

Figure 110c shows a **pressure-equalizing addition funnel**. See that sidearm? Remember when you were warned to remove the stopper of a separatory funnel so that you wouldn't build up a vacuum inside the funnel as you emptied it? Anyway, the sidearm equalizes the pressure on both sides of the liquid you're adding to the flask, so it'll flow freely, without vacuum buildup and without any need for you to remove the stopper. This equipment is very nice, very expensive, very limited, and very rare. And if you try an **extraction** in one of these, all the liquid will run out the tube onto the floor as you shake the funnel.

So a compromise was reached (Fig. 110b). Since you'll probably do more extractions than additions, with or without reflux, the pressure-equalizing tube went out, but the ground-glass joint stayed on. Extractions; no problem. The nature of the stem is unimportant. But during additions, you'll have to take the responsibility to see that nasty vacuum buildup doesn't occur. You can remove the stopper every so often or put a **drying tube** and an **inlet adapter** in place of the stopper. The inlet adapter keeps moisture out and prevents vacuum buildup inside the funnel.

How to Set Up

There are at least two ways to set up an addition and reflux, using either a **three-neck flask** or a **Claisen adapter**. I thought I'd show both of these setups with **drying tubes.** They keep the moisture in the air from getting into your reaction. If you don't need them, do without them.

Often, the question comes up, "If I'm refluxing one chemical, how fast can I add the other reactant?" Try to follow your instructor's suggestions. Anyway, usually the reaction times are fixed. So I'll tell you what NOT, repeat NOT, to do.

If you reflux something, there should be a little ring of condensate, sort of a cloudy, wavy area in the barrel of the reflux condenser (Figs. 111 and 112). Assuming an exothermic reaction—the usual case—adding material from the funnel has the effect of heating up the flask. The ring of condensate begins to move *up*. Well, don't *ever* let this get more than three quarters up the condenser barrel. If the reaction is that fast, a very little extra reagent or heating will push that ring out of the condenser and possibly into the room air. No, no, no, no.

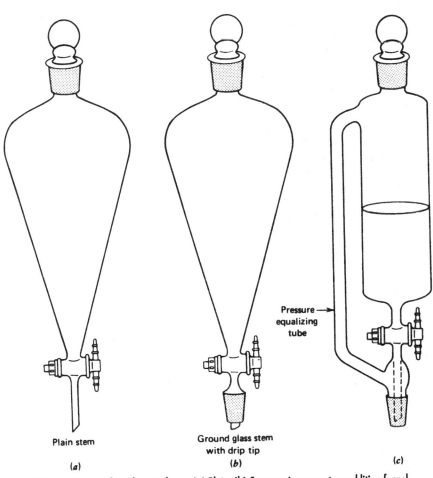

Plain stem

(a)

Ground glass stem
with drip tip

(b)

Pressure →
equalizing
tube

(c)

Fig. 110 Separatory funnels in triplicate. (a) Plain. (b) Compromise separatory addition funnel. (c) Pressure-equalizing addition funnel.

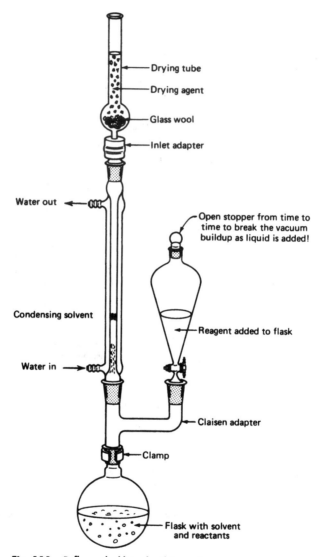

Fig. 111 Reflux and addition by Claisen tube.

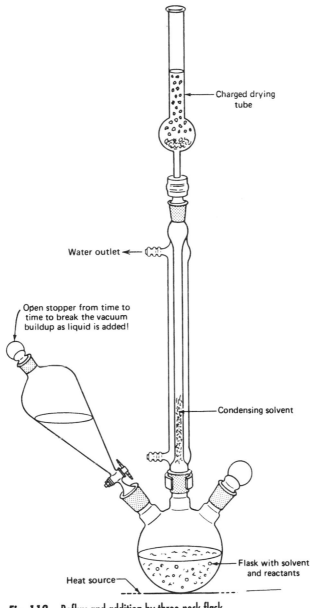

Charged drying tube

Water outlet

Open stopper from time to time to break the vacuum buildup as liquid is added!

Condensing solvent

Flask with solvent and reactants

Heat source

Fig. 112   Reflux and addition by three-neck flask.

23

REFLUX:
MICROSCALE

There's not much difference between the microscale reflux and the bigger ones other than the size of the glassware, so much of the information for the larger setup will still apply (see Chapter 22, "Reflux"). And if you want a dry microscale reflux, then add a drying tube (Fig. 113).

ADDITION AND REFLUX: MICROSCALE

Once you get used to the idea of handling reagents in syringes (see Chapter 8, "Syringes and Needles"), this is pretty straightforward (Fig. 114). Sometimes you have to watch out for back-pressure in the setup, and you have to hold the syringe plunger a bit to keep it from blowing back out. Normally, you just hold the barrel of the loaded syringe, pierce the septum, and then add the reagent, a little at a time, into the setup. How fast? Usually as fast as practical. If the reaction speeds up too much, the ring of condensate in the reflux condenser will travel up and out of the condenser and into the room air. No, no, no, no. This sounds identical to the warning I give for the large-scale setup. **Hint:** It is.

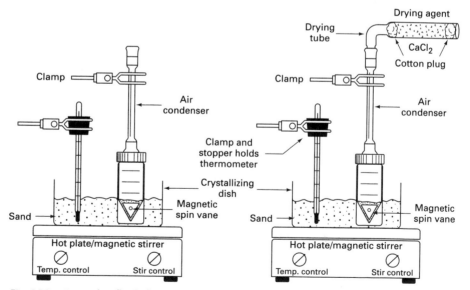

Fig. 113 Microscale reflux both wet and dry.

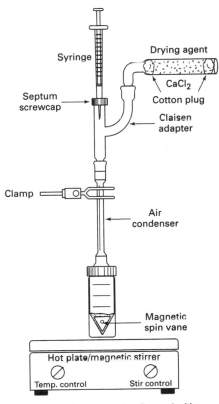

23

Fig. 114 A dry microscale reflux and addition.

SUBLIMATION

Sublimation occurs when you heat a solid and it turns directly into a vapor. It does not pass GO, nor does it turn into a liquid. If you reverse the process—cool the vapor so that it turns back into a solid—you've *condensed the vapor*. Use the unique word, **sublime**, for the direct conversion of *solid to vapor*. **Condense** can refer to either *vapor-to-solid* or *vapor-to-liquid* conversions.

Figure 115 shows three forms of sublimation apparatus. Note all the similarities. Cold water goes in and down into a **cold finger** on which the vapors from the crystals condense. The differences are that one is larger and has a **ground glass joint.** The **sidearm test tube** with **cold-finger condenser** is much smaller. To use them,

1. Put the crude solid into the bottom of the sublimator. How much crude solid? This is rather tricky. You certainly don't want to start with so much that it touches the cold finger. And since, as the purified solid condenses on the cold finger, it begins to grow down to touch the crude solid, there has to be quite a bit of room. I suggest that you see your instructor, who may want only a small amount purified.

2. Put the cold finger into the bottom of the sublimator. Don't let the clean cold finger touch the crude solid. If you have the sublimator with the ground-glass joint, lightly (and I mean lightly) grease the joint. Remember that greased glass joints should NOT be clear all the way down the joint.

3. Attach the hoses. *Cold water goes in the center tube,* pushing the warmer water out the side tube. Start the cooling water. Be careful! The inexpensive microscale sublimator needs no running cooling water—only ice and water in the tube.

4. If you're going to pull a vacuum in the sublimator, do it now. If the vacuum source is a **water aspirator**, *put a water trap between the aspirator and the sublimator.* Otherwise, you may get depressed if, during a sudden pressure drop, water backs up and fills your sublimator. Also, start the vacuum *slowly.* If not, air, entrained in your solid, comes rushing out and blows the crude product all over the sublimator, like popcorn.

5. When everything has settled down, slowly begin to heat the bottom of the sublimator, if necessary. You might see vapors coming off the solid. Eventually, you'll see crystals of *purified solid form on the cold finger.* Since you'll work with different substances, different methods of heating will have to be used. Ask your instructor.

6. Now the tricky parts. You've let the sublimator cool. If you've a vacuum in the sublimator, carefully—very carefully—introduce air into the device. A sudden inrush of air, and PLOP! Your purified crystals are just so much yesterday's leftovers. Start again.

7. Now again, carefully—very carefully—remove the cold finger, with your pristine product clinging tenuously to the smooth glass surface, without a lot of bonking and shaking. Otherwise PLOP! et cetera, et cetera, et cetera. Clean up and start again.

24

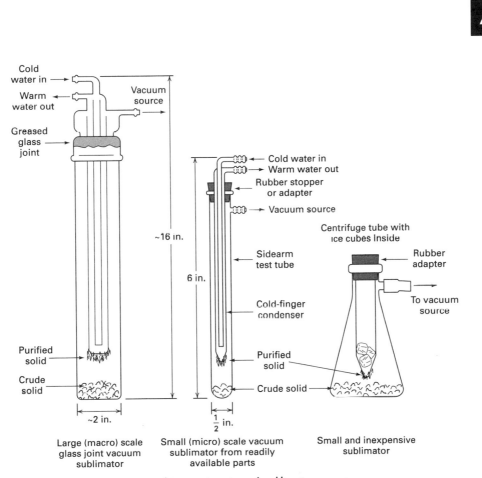

Fig. 115 King-size, miniature, and inexpensive microscale sublimation apparatus.

CHROMATOGRAPHY: SOME GENERALITIES

Chromatography is perhaps the most useful means of separating compounds to purify and identify them. Indeed, separations of colored compounds on paper strips gave the technique its colorful name. Although there are many different types of chromatography, all the forms bear tremendously striking similarities. **Thin-layer, wet-column,** and **dry-column chromatography** are common techniques you'll run across.

This chromatography works by differences in **polarity**. (That's not strictly true for all types of chromatography, but I don't have the inclination to do a 350-page dissertation on the subject, when all you might need to do is separate the differently colored inks in a black marker pen.)

ADSORBENTS

The first thing you need is an **adsorbent**, a porous material that can suck up liquids and solutions. Paper, silica gel, alumina (ultrafine aluminum oxide), corn starch, and kitty litter (unused) are all fine adsorbents. Only the first *three* are used for chromatography, however. You may or may not need a **solid support** with these. Paper hangs together, is fairly stiff, and can stand up by itself. Silica gel, alumina, corn starch, and kitty litter are more or less powders and will need a solid support to hold them.

Now you have an *adsorbent on some support*, or a *self-supporting adsorbent*, like a strip of paper. You also have a mixture of stuff you want to separate. So you dissolve the mixture in an easily evaporated solvent, like methylene chloride, and put some of it on the adsorbent. *Zap!* It is adsorbed! Stuck on and held to the adsorbent. But because you have a mixture of different things, and they are different, they will be *held to the adsorbent in differing degrees*.

SEPARATION OR DEVELOPMENT

Well, now there's this mixture, sitting on this adsorbent, looking at you. Now you start to *run solvents through the adsorbent*. Study the following list of solvents. Chromatographers call these solvents **eluents**.

THE ELUATROPIC SERIES

Not at all like the World Series, **the eluatropic series** is simply a list of solvents arranged according to increasing polarity.

Some solvents arranged in order of increasing polarity

(Least polar)	Pet. ether
	Cyclohexane
Increasing	Toluene
Polarity	Chloroform
	Acetone
	Ethanol
(Most polar)	Methanol

So you start running "pet. ether" (remember, petroleum ether, a mixture of hydrocarbons like gasoline—not a true ether at all). It's not very polar. So it is *not held strongly to the adsorbent.*

Well, this solvent is traveling through the adsorbent, minding its own business, when it encounters the mixture placed there earlier. It tries to kick the mixture out of the way. But most of *the mixture is more polar, held more strongly* on the adsorbent. Since the pet. ether cannot kick out the compounds that are more polar than itself very well, most of *the mixture is left right where you put it.*

No separation

Desperate, you try methanol, one of the most polar solvents. It is *really held strongly to the adsorbent.* So it comes along and kicks the living daylights out of just about *all* the molecules in the mixture. After all, the methyl alcohol *is more polar, so it can move right along and displace the other molecules.* And it does. So, when you evaporate the methanol and look, *all the mixture has moved with the methanol,* so you get *one spot* that moved, right with the **solvent front**.

No separation

Taking a more reasonable stand, you try chloroform, because it has an intermediate polarity. The chloroform comes along, sees the mixture, and is able to push out, say, *all but one* of the components. As it travels, kicking the rest along, it gets tired and starts to leave some of the more polar components behind. After a while, *only one component is left moving with the chloroform*, and that may be dropped, too. So, at the end, *there are several spots left*, and each of them is in a *different place* from the start. Each spot is *at least one different component* of the entire mixture.

Separation. At last!

I picked these solvents for illustration. They are quite commonly used in this technique. I worry about the hazards of using chloroform, however, because it's been implicated in certain cancers. Many other common solvents, also, are suspected to be carcinogens. In lab, either you will be told what solvent (eluent) to use or you will have to find out yourself, mostly by trial and error.

THIN-LAYER CHROMATOGRAPHY: TLC

26

Thin-layer chromatography (TLC) is used for identifying compounds and determining their purity. The most common adsorbent used is **silica gel**, but **Alumina** is gaining popularity, and with good reason. Compounds should separate the same on an alumina plate as on an alumina column, and column chromatography using alumina is still very popular. It is very easy to run test separations on **TLC plates** rather than on **chromatographic columns**.

Nonetheless, both of these adsorbents are powdered and require a **solid support**. **Microscope slides** are extremely convenient. To keep the powder from just falling off the slides, manufacturers add a *gypsum binder* (plaster). Adsorbents with the binder usually have a **"G"** stuck on the name or say "For thin-layer use" on the container.

Sometimes a fluorescent powder is put into the adsorbent to help with **visualization** later. The powder usually glows a bright green when you expose it to **254-nm wavelength ultraviolet (UV) light**. You can probably figure out that if a container of silica gel is labeled **Silica Gel G-254**, you've got a TLC adsorbent with all the bells and whistles.

Briefly, you mix the adsorbent with water, spread the mix on the microscope slide in a *thin layer*, let it dry, and then activate the coating by heating the coated slide on a hot plate. Next you **spot** or place your unknown compound on the plate, let an **eluent** run through the adsorbent (**development**), and finally examine the plate (**visualization**).

PREPARATION OF TLC PLATES

1. Clean and dry several microscope slides.
2. In an Erlenmeyer flask, weigh out some adsorbent and add water.

   a. For silica gel use a 1:2 ratio of gel to water. About 2.5 g gel and 5 g water will do for a start.

   b. For alumina, use a 1:1 ratio of alumina to water. About 2.5 g alumina and 2.5 g water will be a good start.

3. Stopper the flask and shake it until all the powder is wet. This material must be used quickly because a gypsum (plaster) binder is present.
4. Spread the mix by using a *medicine dropper. Do not use disposable pipets!* The disposable pipets have *extremely* narrow openings at the end, and they clog up easily. There exists a "dipping method" for preparing TLC slides, but since the usual solvents, methanol and

chloroform (***Caution!*** *Toxic!*) *do not activate the binder, the powder falls off the plate. Because the layers formed by this process are very thin, they are very fragile.*

5. *Run a bead of mix around the outside of the slide, and then fill the remaining clear space. Leave ¼ inch of the slide blank on one end*, so you can hold onto the slide. *Immediately* tap the slide from the bottom to smooth the mix out (Fig. 116). Repeat this procedure with as many slides as you can. If the mix sets up and becomes unmanageable, add a little water and shake well.

6. Let the slides sit until the gloss of water on the surface has gone. Then place the slides on a hot plate until they dry.

(Caution! If the hot plate is too hot, the water will quickly turn to steam and blow the adsorbent off the slides.)

THE PLATE SPOTTER

1. The **spotter** is the apparatus used to put the solutions you want to analyze on the plate. You use it to make a spot of sample on the plate.

2. Put the center of a **melting point capillary** into a small, blue Bunsen burner flame. Hold it there until the tube softens and starts to sag. Do not rotate the tube, ever.

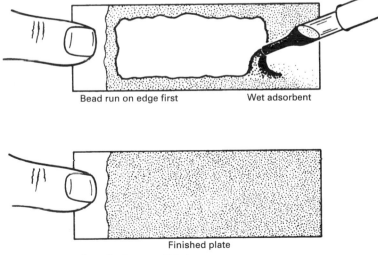

Bead run on edge first   Wet adsorbent

Finished plate

Fig. 116 Spreading adsorbent on a TLC plate.

3. *Quickly* remove the capillary from the flame, and pull both ends (Fig. 117). If you leave the capillary in the flame too long, you get an obscene-looking mess.

4. Break the capillary at the places shown in Fig. 117 to get **two spotters** that look roughly alike. (If you've used capillaries with *both ends open already*, then you don't have a closed end to break off.)

5. Make up 20 of these, or more. You'll need them.

6. Because TLC is so sensitive, spotters tend to "remember" old samples if you reuse them. *Don't put different samples in the same spotter.*

SPOTTING THE PLATES

1. Dissolve a small portion (1–3 mg) of the substance you want to chromatograph in *any* solvent that dissolves it *and* evaporates rapidly. Dichloromethane or diethyl ether often works best.

2. Put the thin end of the capillary spotter into the solution. The solution should rise up into the capillary.

3. Touch the capillary to the plate *briefly*! The compound will run out and form a small spot. Try to keep the spot as small as possible—not larger than ¼ inch in diameter. Blow gently on the spot to evaporate the solvent. Touch the capillary to the *same place*. Let this new spot grow to be *almost the same size as the one already there*. Remove the capillary and gently blow away the solvent. This will build up a concentration of the compound.

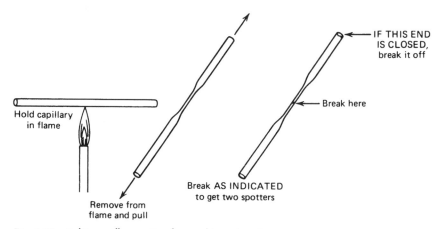

Fig. 117 Making capillary spotters from melting point tubes.

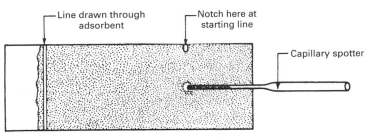

Fig. 118 Putting a spot of compound on a TLC plate.

4. Take a sharp object (an old pen point, capillary tube, spatula edge, etc.) and *draw a straight line through the adsorbent*, as close to the clear glass end as possible (Fig. 118). Make sure that it runs all the way across the end of the slide and goes right down to the glass. This will keep the solvent from running up to the ragged edge of the adsorbent. It will travel only as far as the smooth line you have drawn. Measurements will be taken from this line.

5. Now make a small notch in the plate at the level of the spots to mark their starting position. You'll need this later for measurements.

DEVELOPING A PLATE

1. Take a 150 ml beaker, line the sides of it with filter paper, and cover with a watch glass (Fig. 119). Yes, you can use other containers. I've used beakers because all chemistry stockrooms have beakers, you know what beakers look like, and it's a handy single word for the more

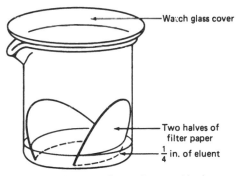

Fig. 119 The secret identity of a 150 ml beaker as a TLC slide development chamber is exposed.

cumbersome "thin-layer development chamber used at your institution." So if you do use jelly jars, drinking glasses, wide-mouth screwcap bottles or whatever, just substitute that description for beaker in this discussion, OK?

2. Choose a solvent to **develop** the plate. You let this solvent (eluent) pass through the adsorbent by capillary action. Nonpolar eluents (solvents) will force nonpolar compounds to the top of the plate, whereas polar eluents will force *both polar and nonpolar materials up the plate*. There is only one way to choose eluents. Educated guesswork. Use the chart of eluents presented in Chapter 25.

3. Pour some of the eluent (solvent) into the beaker, and tilt the beaker so that the solvent wets the filter paper. Put no more than ¼ inch of eluent in the bottom of the beaker! This saturates the air in the beaker with eluent (solvent) and stops the evaporation of eluent from the plate.

4. Place the slide into the developing chamber as shown (Fig. 120). *Don't let the solvent in the beaker touch the spot on the plate,* or the spot will dissolve away into the solvent! If this happens, you'll need a new plate, and you'll have to clean the developing chamber as well.

5. Cover the beaker with a watch glass. The solvent (eluent) will travel up the plate. The filter paper keeps the air in the beaker saturated with solvent so that it doesn't evaporate from the plate. When the solvent reaches the line, *immediately* remove the plate. Drain the solvent from it, and blow gently on the plate until *all* the solvent is gone. If not, there will be some trouble visualizing the spots.

Don't breathe the fumes of the eluents! Make sure you have adequate ventilation. Work in a hood if possible.

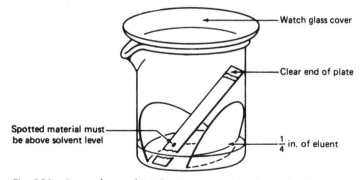

Fig. 120 Prepared, spotted TLC plate in a prepared developing chamber.

VISUALIZATION

Unless the compound is colored, the plate will be blank and you won't be able to see anything, so *you must visualize the plate*.

1. **Destructive visualization.** Spray the plate with sulfuric acid, then bake in an oven at 110°C for 15–20 minutes. Any spots of compound will be charred blots, utterly destroyed. *All* spots of compound will be shown.

2. **Semidestructive visualization.** Set up a developing tank (150 ml beaker) but leave out the filter paper and any solvent. Just a beaker with a cover. Add a few crystals of **iodine**. Iodine vapors will be absorbed onto most spots of compound, coloring them. Removing the plate from the chamber causes the iodine to evaporate from the plate, and *the spots will slowly disappear. Not all spots may be visible.* So if there's nothing there, that doesn't mean nothing's there. The iodine might have reacted with some spots, changing their composition. Hence the name **semidestructive visualization.**

3. **Nondestructive visualization.**

   a. Long-wave UV (Hazard!) Most TLC adsorbents contain a fluorescent powder that glows bright green when you put them under long-wave UV light. There are two ways to see the spots:

   (1) The background glows green, and the spots are dark,

   (2) The background glows green, and the spots glow some other color. The presence of *excess eluent may cause whole sections of the plate to remain dark.* Let all the eluent evaporate from the plate.

   b. Short-wave UV (Hazard!) The plates stay dark. *Only the compounds may glow.* This is usually at 180 nm.

Both of the UV tests can be done in a **UV light box** in a matter of seconds. Since most compounds are unchanged by exposure to UV, the test is considered **nondestructive**. Not everything will show up, but the procedure is good enough for most compounds. When using the light box, *always turn it off when you leave it.* If you don't, not only does the UV filter burn out, but also your instructor becomes displeased.

Since neither the UV nor the iodine test is permanent, it helps to have a record of what you've seen. You must *draw an accurate picture of the plate in your notebook*. Using a sharp-pointed object (pen point, capillary tube, etc.), you can trace the outline of the spots on the plate while they are under the UV light (**Caution!** Wear gloves!) or before the iodine fades from the plate.

INTERPRETATION

After visualization, there will be a spot or spots on the plate. Here is what you do when you look at them.

1. Measure the distance *from that solvent line* drawn across the plate *to where the spot started*.
2. Measure the distance from where the *spot stopped* to where the spot began. Measure to the *center of the spot* rather than to one edge. If you have more than one spot, get a distance for each. If the spots have funny shapes, do your best.
3. Divide the *distance the solvent moved* into the *distance the spot(s) moved*. The resulting ratio is called the R_f **value**. *Mathematically*, the ratio for any spot should be between 0.0 and 1.0, or you goofed. *Practically*, spots with R_f values greater than about 0.8 and less than about 0.2 are hard to interpret. They could be single spots or multiple spots all bunched up and hiding behind one another.
4. Check out the R_f **value**—it may be helpful. In identical circumstances, this value would always be the *same* for a single compound, all the time. If this were true, you could identify unknowns by running a plate and looking up the R_f value. Unfortunately, the technique is not that good, but you can use it *with some judgment* and a *reference compound* to identify unknowns (see "Multiple Spotting," following).

Figures 121 to 123 provide some illustrations. Look at Fig. 121: if you had a mixture of compounds, you could never tell. This R_f value gives no information. Run this compound again. Run a new plate. *Never redevelop an old plate!*

Use a more polar solvent!

No information in Fig. 122 either. You couldn't see a mixture if it were here. Run a new plate. *Never redevelop an old plate!*

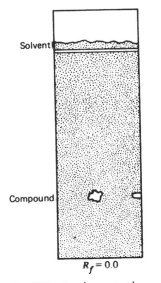

$R_f = 0.0$

Fig. 121 Development with a nonpolar solvent and no usable results.

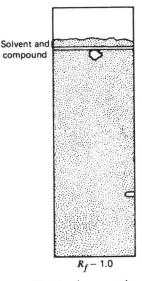

$R_f - 1.0$

Fig. 122 Development with a very polar solvent and no usable results.

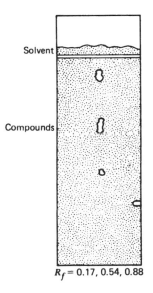

$R_f = 0.17, 0.54, 0.88$

Fig. 123 Development with just the right solvent is a success.

Use a less polar solvent!

If the spot moves somewhere between the two limits (shown in Fig. 123) and *remains a single spot*, the compound is pure. If *more than one spot shows*, the compound is impure and it is a **mixture**. Whether the compound should be purified is a matter of judgment.

MULTIPLE SPOTTING

You can run more than one spot, either to save time or to make comparisons. You can even *identify unknowns*.

Let's say that there are two unknowns, A and B. Say one of them can be biphenyl (a colorless compound that smells like moth balls). You spot two plates: One with A and biphenyl, side by side, and the other, B and biphenyl, side by side. After you develop both plates, you have the results shown in Fig. 124.

Apparently, A is biphenyl.

Note that the R_f values are not perfect. This is an imperfect world, so don't panic over a slight difference.

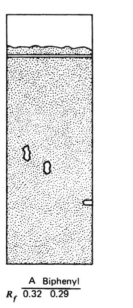
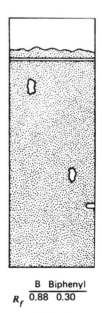

	A	Biphenyl		B	Biphenyl
R_f	0.32	0.29	R_f	0.88	0.30

Fig. 124 Side-by-side comparison of an unknown and a leading brand known.

And now we have a method that can quickly determine

1. Whether a compound is a mixture.
2. The identity of a compound *if a standard is available.*

There are **pre-prepared plates** that have an active coating on a thin plastic sheet, also with or without the fluorescent indicator. You can cut these to any size (they are about 8 inches by 8 inches) with a pair of scissors. *Don't touch the active surface with your fingers—handle them only by the edge.* The layers on the plate are much thinner than those you would make by spreading adsorbent on a microscope slide, so you have to use smaller amounts of your compounds in order not to overload the adsorbent.

CO-SPOTTING

You can nail down identification with **co-spotting**, but it is a bit tricky. This time, spot one plate with your unknown, A, in two places. Now let the spots dry entirely. I mean entirely! Now we'll spot biphenyl *right on top of one of the spots of A.* (If you didn't let the earlier spot dry entirely, the biphenyl spot would bleed and you'd have a *very* large—and useless—spot of mixed compounds.) Do the same for B on another plate. Now run these plates.

The R_f values might be a bit different, but there's only *one spot* in the biphenyl over the A column. No separation. They are the same. If you see two (or more) spots come from co-spotting, the substances, like B, are not the same (Fig. 125).

OTHER TLC PROBLEMS

What you've seen is what you get if everything works OK. You can:

1. Make the spots too concentrated. Here spots smear out, and you don't see any separation. Dilute the spotting solution or don't put so much on (Fig. 126).
2. Put the spots too close together. Here you could get the spots bleeding into each other, and you might not be able to tell which spot came from which origin (Fig. 127).

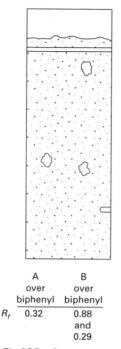

	A over biphenyl	B over biphenyl
R_f	0.32	0.88 and 0.29

Fig.125 Co-spotting
biphenyl over A and B.

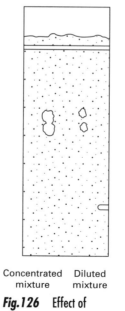

Concentrated   Diluted
mixture   mixture

Fig.126 Effect of
concentration on separation.

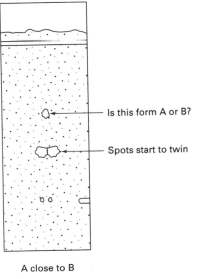
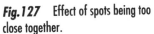

Is this form A or B?

Spots start to twin

A close to B

Fig. 127 Effect of spots being too close together.

26

3. Spot too close to the edge. Here you get inaccurate R_f values, bleeding, and other problems. The spots are not surrounded as much by adsorbent and solvent, so unequal forces are at work here (Fig. 128).

Fig. 128 Effect of spot too close to edge.

PREPARATIVE TLC

When you use an **analytical technique** (like TLC) and you expect to **isolate compounds**, it's often called a **preparative (prep) technique**. So TLC becomes **Prep TLC**. You use the same methods, only on a larger scale.

Instead of a microscope slide, you usually use a 12 X 12 inch glass plate and coat it with a thick layer of adsorbent (0.5-2.0 mm). Years ago, I used a small paintbrush to put a line (a streak rather than a spot) across the plate near the bottom. Now you can get special plate streakers that give a finer line and less spreading. You put the plate in a large developing chamber and develop and visualize the plate as usual.

The thin line separates and spreads into bands of compounds, much like a tiny spot separates and spreads on the analytical TLC plates. Rather than just look at the bands, though, you *scrape the adsorbent holding the different flasks*, blast your compounds off of the adsorbents with appropriate solvents, filter off the adsorbent, and finally evaporate the solvents and actually recover the separate compounds.

WET-COLUMN
CHROMATOGRAPHY

27

Wet-column chromatography, as you may have guessed, is chromatography carried out on a **column of adsorbent**, rather than a layer. Not only is it cheap, easy, and carried out at room temperature but also you can separate large amounts, *gram quantities*, of mixtures.

In column chromatography, the adsorbent is either alumina or silica gel. Alumina tends to be basic and silica gel tends to be acidic. If you try out an eluent (solvent) on silica gel plates, you should use a silica gel adsorbent. And if you have good results on alumina TLC, use an alumina column.

Now you have a *glass tube as the support* holding the adsorbent in place. You dissolve your mixture and put it on the adsorbent at the top of the column. Then you wash the mixture down the column using at least one eluent (solvent), perhaps more. The compounds carried along by the solvent are washed *entirely out of the column*, into separate flasks. Then you isolate the separate fractions.

PREPARING THE COLUMN

1. The adsobent is supported by a glass tube with either a stopcock or a piece of tubing and a screw clamp to control the flow of eluent (Fig. 129). You can use an ordinary buret. What you will use will depend on your own lab program. Right above this control you put a wad of cotton or glass wool to keep everything from falling out. *Do not use too much cotton or glass wool*, and *do not pack it too tightly*. If you ram the wool into the tube, the flow of eluent will be very slow, and you'll be in lab until next Christmas waiting for the eluent. If you pack it *too loosely*, all the stuff in the column *will fall out*.

2. At this point, fill the column half-full with the least polar eluent you will use. If this is not given, you can guess it from a quick check of separation of the mixture on a TLC plate. This would be the advantage of an alumina TLC plate.

3. Slowly put sand into the column through a funnel until there is a 1 cm layer of sand over the cotton. Adsorbent alumina or silica gel is so fine that either is likely to go through cotton but not through a layer of fine sand.

4. During this entire procedure, *keep the level of the solvent above that of any solid material in the column*!

5. Now slowly add the adsorbent. Alumina or silica gel are adsorbents and they'll suck up the solvent. When they do, heat is liberated. The solvent may boil and *ruin the column*. Add the adsorbent slowly! Use

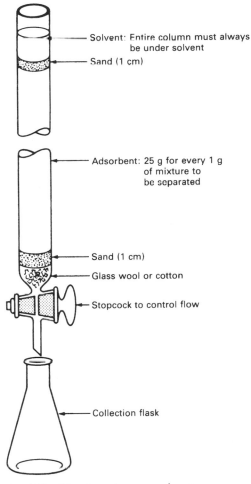

Solvent: Entire column must always be under solvent

Sand (1 cm)

Adsorbent: 25 g for every 1 g of mixture to be separated

Sand (1 cm)

Glass wool or cotton

Stopcock to control flow

Collection flask

Fig. 129 Wet-column chromatography setup.

about 25 g of adsorbent for every 1 g of mixture you want to separate. While adding the adsorbent, *tap or gently swirl the column to dislodge any adsorbent or sand on the sides.* You know, a plastic wash bottle with eluent in it can wash the stuff down the sides of the column very easily.

6. When the alumina or silica gel settles, you normally have to add sand (about 1 cm) to the top to keep the adsorbent from moving around.

7. Open the stopcock or clamp and let solvent out until the level of the solvent is just above the upper level of sand.

8. *Check the column! If there are air bubbles or cracks in the column, dismantle the whole business and start over!*

COMPOUNDS ON THE COLUMN

If you've gotten this far, congratulations! Now you have to get your mixture, the analyate, on the column. Dissolve your mixture in the same solvent you are going to put through the column. Try to keep the volume of the solution of mixture as *small as possible*. If your mixture does not dissolve entirely *(and it is important that it do so)*, check with your instructor! You might be able to use different solvents for the analyate and for the column, but this isn't as good. You might use the *least polar* solvent that will dissolve your compound.

If you must use the column eluent as the solvent, and not all the compound will dissolve, you can filter the mixture through filter paper. Try to keep the volume of solution down to 10 ml or so. After this, the sample becomes unmanageable.

1. Use a pipet and rubber bulb to slowly and carefully add it to the top of the column (Fig. 130). *Do not disturb the sand!*
2. Open the stopcock or clamp and let solvent flow out until the level of the solution of compound is slightly above the sand. *At no time let the solvent level get below the top layer of sand!* The compound is now "on the column."

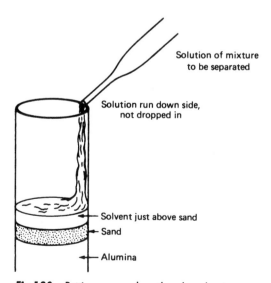

Solution of mixture
to be separated

Solution run down side,
not dropped in

Solvent just above sand

Sand

Alumina

Fig 130 Putting compounds on the column by pipet.

3. Now add eluent (solvent) to the column above the sand. *Do not disturb the sand!* Open the stopcock or clamp. Slowly let eluent run through the column until the first compound comes out. Collect the different products in Erlenmeyer flasks. You may need lots and lots of Erlenmeyer flasks. *At no time let the level of the solvent get below the top of the sand!* If necessary, stop the flow, add more eluent, and start the flow again.

VISUALIZATION AND COLLECTION

If the compounds are colored, you can watch them travel down the column and separate. If one or all are colorless, you have problems. So:

1. *Occasionally* let one or two drops of eluent fall on a clean glass microscope slide. Evaporate the solvent and see if there is any sign of your crystalline compound! This is an excellent spot test, but don't be confused by nasty plasticizers from the tubing trying to put one over on you, pretending to be your product.

2. Put the narrow end of a "TLC spotter" to a drop coming off at the column. The drop will rise up into the tube. Using this loaded spotter, *spot, develop, and visualize a TLC plate with it.* Not only is this more sensitive, but also you can see whether the stuff coming out of the column is pure (see Chapter 26, "Thin-Layer Chromatography"). You'll probably have to collect more than one drop on a TLC plate. If it is very dilute, the plate will show nothing, even if there actually is compound there. It is best to sample four or five consecutive drops.

Once the first compound or compounds have come out of the column, those that are left may move down the column much too slowly for practical purposes. Normally you start with a nonpolar solvent. But by the time all the compounds have come off, it may be time to pick up your degree. The solvent may be too nonpolar to kick out later fractions. So you have to decide to change to a more polar solvent. This will kick the compound right out of the column.

To change solvent in the middle of a run:

1. Let the old solvent level run down to just above the top of the sand.
2. *Slowly add new,* **more polar solvent** *and do not disturb the sand.*

You, and you alone, have to decide if and when to change to a more polar solvent. (Happily, sometimes you'll be told.)

If you have only two components, start with a nonpolar solvent, and when you are sure the *first component is completely off the column, change to a really polar one*. With only two components, it doesn't matter what polarity solvent you use to get the second compound off the column.

Sometimes the solvent evaporates quickly and leaves behind a "fuzz" of crystals around the tip (Fig. 131). Just use some fresh solvent to wash them down into the collection flask.

Now all the components are off the column and in different flasks. Evaporate the solvents (*No flames!*), and lo! The crystalline material is left.

Dismantle the column. Clean up. Go home.

WET-COLUMN CHROMATOGRAPHY: MICROSCALE

Occasionally, you'll be asked to clean and/or dry your product using **column chromatography** by pipet. It's a lot like large-scale wet-column chromatography.

One problem with this pipet column is that there's no stopcock. Once it starts running, it goes until it runs out. And you can't let the column dry out either. So it looks like you're committed to running the column once you prepare it.

In general (Fig. 132):

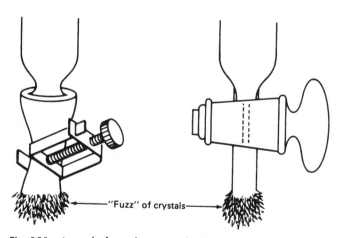

Fig. 131 A growth of crystals occurs as the eluent evaporates.

1. Shorten the tip of a Pasteur pipet (see "Pipet Cutting" in Chapter 7, "Pipet Tips") and tamp cotton into the tip as usual.
2. Add about 50 mg of fine sand.
3. Now comes the adsorbent, about 500 mg or so. Gently tap the pipet to pack this down.
4. Put a bit more sand on top to keep the adsorbent from flying around.
5. Now slowly add the solvent and let it wet the entire column.
6. Dissolve your sample in a minimum amount of solvent—the least polar solvent it'll dissolve in—and add this solution, slowly, to the top of the pipet. Don't fill the pipet with liquid; drip the solution onto the sand.
7. With the sample all on the column, just below the sand, add a small column of **elution solvent**. This may or may not be different from the solvent you dissolved your sample in.
8. Wait and watch. Watch and wait. Do not ever let the column run dry. Collect the drops off the end in a series of fractions in some small containers.

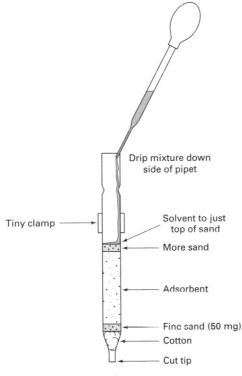

Fig. 132 Pasteur pipet wet-column chromatography.

REFRACTOMETRY

When light travels from one medium to another, it changes velocity and direction a bit. If you've ever looked at a spoon in a glass of water, the image of the spoon in water is displaced a bit from the image of the spoon in air, and the spoon looks broken. When the light rays travel from the spoon in water and break out into the air, they are **refracted**, or shifted (Fig. 133). If we take the ratio of the sine of the angles formed when a light ray travels from air to water, we get a single number, the **index of refraction**, or **refractive index**. Because we can measure the index of refraction to a few parts in 10,000, this is a very accurate physical constant for identification of a compound.

The refractive index is usually reported as ⬚, where the tiny 25 is the temperature at which the measurement was taken, and the tiny capital D means we've used light from a sodium lamp, specifically a single yellow frequency called the sodium D line. Fortunately, you don't have to use a sodium lamp if you have an Abbé refractometer.

THE ABBÉ REFRACTOMETER(FIG. 134)

The refractometer looks a bit like a microscope. It has

1. ***An eyepiece.*** You look in here to make your adjustments and read the refractive index.

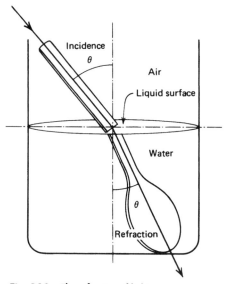

Fig. 133 The refraction of light.

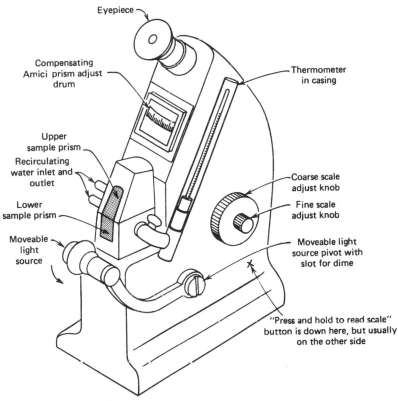

Eyepiece

Compensating
Amici prism adjust
drum

Thermometer
in casing

Upper
sample prism

Recirculating
water inlet and
outlet

Coarse scale
adjust knob

Lower
sample prism

Fine scale
adjust knob

Moveable
light
source

Moveable light
source pivot with
slot for dime

"Press and hold to read scale"
button is down here, but usually
on the other side

Fig. 134 One model of a refractometer.

2. *A compensation prism adjustment*. Since the Abbé refractometer uses white light and not light of one wavelength (the sodium D line), the white light *disperses* as it goes through the optics, and rainbow-like color fringing shows up when you examine your sample. By turning this control, you rotate some compensation prisms that eliminate this effect.

3. *Hinged sample prisms*. This is where you put your sample.

4. *Light source*. This provides light for your sample. It's on a movable arm, so you can swing it out of the way when you place your samples on the prisms.

5. *Light source swivel arm lock*. This is a large, slotted nut that works itself loose as you move the light source up and down a few times. Always have a dime handy to help you tighten this locking nut when it gets loose.

6. *Sample and scale image adjust.* You use this knob to adjust the optics so that you see a split field in the eyepiece. The refractive index scale also moves when you turn this knob. The knob is often a dual control; use the other knob for a coarse adjustment and the inner knob as a fine adjustment.

7. *Scale/sample field switch.* Press this switch, and the numbered refractive index scale appears in the eyepiece. Release this switch, and you see your sample in the eyepiece. Some models don't have this type of switch. You have to change your angle of view (shift your head a bit) to see the field with the refractive index reading.

8. *Line cord on-off switch.* This turns the refractometer light source on and off.

9. *Water inlet and outlet.* These are often connected to temperature-controlled water recirculating baths. The prisms and your samples in the prisms can all be kept at the temperature of the water.

USING THE ABBÉ REFRACTOMETER

1. Make sure the unit is plugged in. Then turn the on-off switch to ON. The light at the end of the movable arm should come on.

2. Open the hinged sample prisms. Not *touching the prisms at all*, place a few drops of your liquid on the lower prism. Then swing the upper prism back over the lower one and *gently* close the prisms. *Never touch the prisms with any hard object, or you'll scratch them.*

3. Raise the light on the end of the movable arm so that the light illuminates the upper prism. Get out your dime and, with the permission of your instructor, tighten the light source swivel arm locknut as it gets tired and lets the light drop.

4. Look in the eyepiece. Slowly, carefully, with *very little* force, turn the large scale and sample image adjust knob from one end of its rotation to the other. *Do not force!* (If your sample is supposedly the same as that of the last person to use the refractometer, you shouldn't have to adjust this much, if at all.)

5. Look for a split optical field of light and dark (Fig. 135). This may not be very distinct. You may have to raise or lower the light source and scan the sample a few times.

6. If you see color fringing at the boundary between light and dark (usually red or blue), slowly turn the compensating prism adjust until

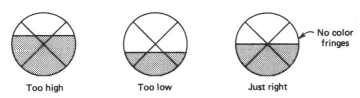

Fig. 135 Your sample through the lens of the refractometer.

the colors are at a minimum. After you do this, you may have to go back and readjust the sample image knob a bit.

7. Press and hold the scale/sample field switch. The refractive index scale should appear (Fig. 136). Read the uppermost scale, the refractive index, to four decimal places. (If your model has two fields, with the refractive index *always* visible, just read it.)

REFRACTOMETRY HINTS

1. The refractive index changes with temperature. If your reading is not the same as that of a handbook, check the temperatures before you despair.
2. Volatile samples require quick action. Cyclohexene, for example, has been known to evaporate from the prisms of unthermostatted refractometers more quickly than you can obtain the index. It may take several tries as you readjust the light, turn the scale and sample image adjust, and so on.
3. Make sure the instrument is *level*. Often, organic liquids can seep out of the jaws before you are ready to make your measurement.

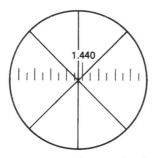

Fig. 136 A refractive index of 1.4398.

INSTRUMENTATION
IN THE LAB

29

Electronic instrumentation is becoming more and more common in the organic lab, a change that is both good and bad. The good part is that you'll be able to analyze your products, or unknowns, much faster, and potentially with more accuracy than ever. The bad part is that you have to learn about how to use the instrumentation, and there are many different manufacturers of different models of the same instrument.

Even worse in this respect is the introduction of computer-controlled instrumentation in the laboratory. Assuming the computer that drives the instrument is an IBM-compatible (many aren't), which compatible is it? An XT? A 286 AT? A 386? A DOS or OS/2 operating system? And then there's the software that drives the instrument. Is it all mouse-driven? What keystrokes mean what? I use a Mattson Galaxy model 2020 **Fourier transform infrared (FTIR)** instrument. I'm familiar with the Perkin-Elmer model, too. These two models look different, their computers are different, and their software is *really* different. I can't justify to myself the inclusion of one model over another because they are so different. (Yes, I have a Perkin-Elmer 710B as an example of a dispersive instrument, but dispersive instruments have more things in common than FTIR's.) So, unless someone has a really good approach and would like to tell me about it, I'll just have to think about it a bit more.

The usual textbook approach is to take a piece of equipment, say something like "This is a typical model," and go on from there, trying to illustrate some very common principles. But what if your equipment is different? Well, that's where you'll have to rely on your instructor to get you out of the woods. I'm going to pick out specific instruments as well. But at least now you won't panic if the knobs and settings on your instrument are not quite the same.

With that said, I'd like to point out a few things about the discussions that follow.

1. If you just submit samples to be run on various instruments, as I did as an undergraduate, pay most attention to the **sample preparation** sections. They say little more than "Don't hand in a dirty sample," but often that's enough.
2. If you get to put the sample into the instruments yourself, **sample introduction** is just for you.
3. If you get to play with (in a nonpejorative sense) the instrument settings, you'll have to wade through the entire description.

You'll notice that I've refrained from calling these precision instruments "machines." That's because they *are* precision instruments, not machines—unless they don't work.

GAS
CHROMATOGRAPHY

Gas chromatography (GC) can also be referred to as **vapor-phase chromatography** (VPC) and even **gas-liquid chromatography** (GLC). Usually, the technique, the instrument, and the chart recording of the data are called **GC:**

"Fire up the GC."	(the instrument)
"Analyze your sample by GC."	(perform the technique)
"Get the data off the GC."	(analyze the chromatogram)

I've mentioned the similarity of all chromatography, and just because electronic instrumentation is used, there's no need to feel that something basically different is going on.

THE MOBILE PHASE: GAS

In **column chromatography** the mobile (moving) phase is a liquid that carries your material through an adsorbent. I called this phase the **eluent**, remember? Here a gas is used to push, or *carry*, your vaporized sample, and it's called the **mobile phase.**

The **carrier gas** is usually helium, although you can use nitrogen. You use a **microliter syringe** to inject your sample into this gas stream through an **injection port**, then *onto the column*. If your sample is a mixture, the *compounds separate on the column* and reach the **detector** at different times. As each component hits the detector, the detector generates **an electric signal.** Usually, the signal goes through an **attenuator network,** and then out to a **chart recorder** to record the signal. I know, it's a fairly general description and Fig. 137 is a highly simplified diagram, but there are lots of different GCs, so being specific about their operation doesn't help here. You should see your instructor. But that doesn't mean we can't talk about some things.

GC SAMPLE PREPARATION

Sample preparation for GC doesn't require much more work than handing in a sample to be graded. *Clean and dry,* right? Try to take care that the boiling point of the material is low enough that you can actually work with the technique. The maximum temperature depends on the type of column,

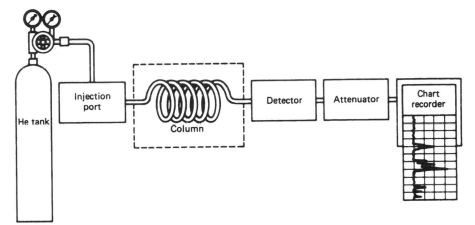

Fig. 137 Schematic of a gas-chromatography setup.

and that should be given. In fact, for any single experiment that uses GC, the nature of the column, the temperature, and most of the electronic settings will be fixed.

GC SAMPLE INTRODUCTION

The sample enters the GC at the **injection port** (Fig. 138). You use a microliter syringe to pierce the rubber septum and inject the sample onto the column. Don't stab yourself or anyone else with the needle. Remember,

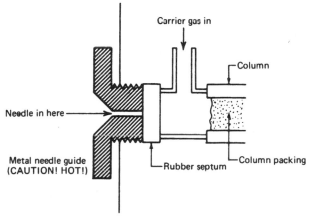

Fig. 138 A GC injection port.

this is *not* dart night at the pub. Don't throw the syringe at the septum. There is a way to do this.

1. To load the sample, put the needle into your liquid sample and *slowly* pull the plunger to draw it up. If you move too fast, and more air than sample gets in, you'll have to push the plunger back again and draw it up once more. Usually, they give you a 10[m>l syringe, and 1 or 2[m>l of sample is enough. Take the loaded syringe out of the sample, and *carefully, cautiously,* pull the plunger back so there is no sample in the needle. You should see a bit of air at the very top but not very much. This way you don't run the risk of having your compound boil out of the needle as it enters the injector oven just before you actually inject your sample. That broadens the sample and reduces the resolution. In addition, the air acts as an internal standard. Since air travels through the column almost as fast as the carrier gas, the **air peak** that you get can signal the start of the chromatogram, much like the notch at the start of a TLC plate. Ask your instructor.
2. Hold the syringe in *two hands* (Fig. 139). There is no reason to practice being an M.D. in the organic laboratory.
3. Bring the syringe to the level of the injection port, straight on. No angles. Then let the needle touch the septum at the center.
4. The real tricky part is holding the barrel and, without injecting, pushing the needle through the septum. This is easier to write about than it is to do the first time.
5. Now quickly and smoothly push on the plunger to inject the sample, and pull the syringe needle out of the septum and injection port.

After a while, the septum gets full of holes and begins to leak. Usually, you can tell you have a leaky septum when the pen on the chart recorder wanders about aimlessly without any sample injected.

SAMPLE IN THE COLUMN

Now that the sample is in the column, you might want to know what happens to this mixture. Did I say mixture? Sure. Just as with **thin-layer** and **column chromatography**, you can use GC to determine the composition or purity of your sample.

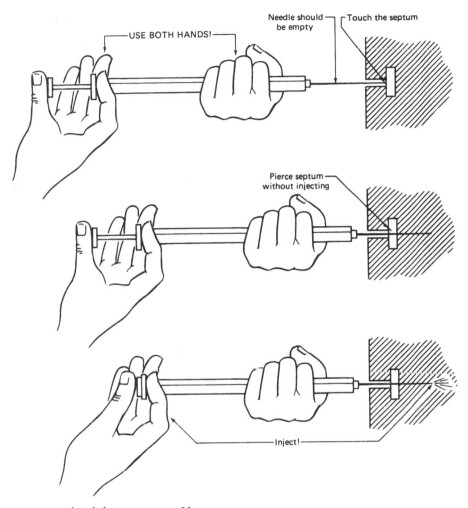

Fig. 139 Three little steps to a great GC.

Let's start with two components, A and B again, and follow their path through an **adsorption column**. Well, if A and B are different, they are going to stick on the adsorbent to different degrees and spend more or less time flying in the carrier gas. Eventually, one will get ahead of the other. Aha! *Separation*—just like column and thin-layer chromatography. Only here the samples are vaporized, and it's called **vapor-phase chromatography (VPC).**

Some of the adsorbents are coated with a **liquid phase.** Most are very high-boiling liquids, and some look like waxes or solids at room temperature. Still, they're liquid phases. So, the different components of the mixture you've injected will spend different amounts of time in the liquid phase and, again, you'll get a separation of components in your compound. Thus the technique is known as **gas-liquid chromatography (GLC).** Thus you could use the same adsorbent and different liquid phases, and change the characteristics of each column. Can you see how the sample components would partition themselves between the gas and liquid phases and separate according to, perhaps, molecular weight, polarity, size, and so on, making this technique also known as **liquid-partition chromatography?**

Since these liquid phases on the adsorbent are eventually liquids, you can boil them. And that's why there are temperature limits for columns. It is not advisable to heat a column past the recommended temperature, boiling the liquid phase right off the adsorbent and right out of the instrument.

High temperature and air (oxygen) are death for some liquid phases, since they oxidize. So make sure the carrier gas is running through them at all times, even a tiny amount, while the column is hot.

SAMPLE AT THE DETECTOR

There are several types of **detectors**, devices that can tell when a sample is passing by them. They detect the presence of a sample and convert it to an electrical signal that's turned into a **GC peak** (Fig. 140) on the chart recorder. The most common type is the **thermal conductivity detector.** Sometimes called hot-wire detectors, these devices are very similar to the filaments you find in light bulbs, and they require some care. Don't ever turn on the **filament current** unless the carrier gas is flowing. A little air (oxygen), a little heat, a little current, and you get a lot of trouble replacing the burned-out detector.

Usually, there are at least two thermal conductivity detectors in the instrument, in a "bridge circuit." Both detectors are set in the gas stream, but only one gets to see the samples. The electric current running through them heats them up, and they lose heat to the carrier gas at the same rate.

As long as no sample, only carrier gas, goes over *both* detectors, the bridge circuit is **balanced.** There is no signal to the recorder, and the pen does not move.

Now a sample in the carrier gas goes by one detector. This sample has a thermal conductivity different from that of pure carrier gas. So the **sample detector** loses heat at a different rate from the **reference detector.**

Height (mm)	8	68	145
Width at half-height (mm)	4	4	4
Area (mm²)	32	212	580
Relative area	1	8.5	18.1
Distance from injection (in.)	3	$3^{5}/_{16}$	$3^{11}/_{16}$
Retention times t_R at 2 min/in. (min)	6	6.63	7.38

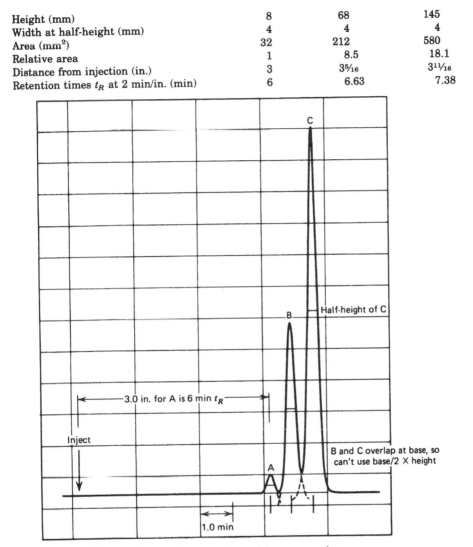

Fig. 140 A well-behaved GC trace showing a mixture of three compounds.

(Remember, the **reference** is the detector that NEVER sees samples—only carrier gas.) The detectors are in different surroundings. They are not really equal any more. So the bridge circuit becomes unbalanced, and a signal goes to the chart recorder, giving a GC peak.

Try to remember the pairing of **sample** with **reference**; it's the *difference* in the two that most electronic instrumentation responds to. You will see this again and again.

ELECTRONIC INTERLUDE

The electrical signal makes two other stops on its way to the chart recorder.

1. *The coarse attenuator.* This control makes the signal weaker (attenuates it). Usually, there's a scale marked in **powers of two**: 2, 4, 8, 16, 32, 64, ... So each position is half as sensitive as the last one. There is one setting, either an [8] or an **S** (for **shorted**), which means that the attenuator has *shorted out the terminals connected to the chart recorder.* Now the **chart recorder zero** can be set properly.
2. *The GC zero control.* This control helps set the zero position on the chart recorder, but it is *not to be confused with the zero control on the chart recorder.*

Here's how to set up the electronics properly for a GC and a chart recorder.

1. The **chart recorder** and **GC** should be allowed to warm up and stabilize for at least 10–15 minutes. Some systems take more time; ask your instructor.
2. Set the **coarse attenuator** to the highest attenuation, usually an [8] or an S.
3. Now set the pen on the chart recorder to zero using *only the chart recorder zero control.* Once you do that, *leave the chart recorder alone.*
4. Start turning the *coarse attenuator* control to more sensitive settings (lower numbers) and *watch the pen on the chart recorder.*
5. If the pen on the chart recorder moves off zero, *use the GC zero control only* to bring the pen back to the zero line on the chart recorder paper.
6. *Do not touch the chart recorder zero. Use the GC zero control only.*
7. As the *coarse attenuator* gets to more sensitive settings (lower numbers), it becomes more difficult to adjust the *chart recorder pen to zero* using *only the GC zero control.* Do the best you can at the *lowest attenuation* (highest sensitivity) you can hold a zero steady at.
8. Now, you *don't* normally run samples on the GC at attenuations of 1 or 2. These settings are *very sensitive,* and there may be lots of electrical noise—the pen jumps about. The point is, *if the GC zero is OK at an attenuation of 1,* then when you run at attenuations of 8, 16, 32, and so on, *the baseline will not jump if you change attenuation in the middle of the run.*

Now that the attenuator is set to give peaks of the proper height, you're ready to go. Just be aware that there may be a **polarity switch** that can make your peaks shift direction.

SAMPLE ON THE CHART RECORDER

Interpreting a GC is about the same as interpreting a TLC plate, so I'll use TLC terms as a comparison to show the similarities. Remember the R_f value from TLC? The ratio of the distance the eluent moved to how far the spots of compound moved? Well, distances can be related to times, so the equivalent of R_f in GC is **retention time.** It's the time it takes the sample to move through the column minus the time it takes for the carrier gas to move through the column. Remember the part about putting air into the syringe to get an **air peak?** Well, you can assume that air travels with the carrier gas and doesn't interact with the column material. So the air peak that shows up on the chart paper can be considered to be the reference point, the "notch," as it were, that marks the start, just as on the TLC plates.

OK, so you don't want to use an air peak. Then make a mark on the chart paper as soon as you've injected the sample, and use *that* as the start. It's not as good, but it'll work.

No. You do not need a stopwatch for the retention times. Find out the distance the chart paper crawls in, say, a minute. Then get out your little ruler and measure the distances from the starting point (either air peak or pen mark) to the midpoint of each peak on the baseline (Fig. 140). Don't be wise and do any funny angles. It won't help. You've got the distances and the chart speed, so you've got the retention time. It works out. Trust me.

You can also estimate how much of each compound is in your sample by measuring **peak areas.** The area under each GC peak is proportional to the amount of material that's come by the detector in that fraction. You might have to make a few assumptions (e.g., the peaks are truly triangular, and each component gives the same response at the detector), but usually it's pretty straightforward. Multiply the height of the peak by the width at half the height. If this sounds suspiciously like the area of a triangle, you're on the right track. It's usually not half the base times the height, however, since sometimes the **baseline** is not very even, and that measurement is difficult.

PARAMETERS, PARAMETERS

To get the best GC trace from a given column, there are lots of things you can do, simply because you have so many controls. Usually, you'll be told the correct conditions, or they'll be preset on the GC.

Gas Flow Rate

The faster the carrier gas flows, the faster the compounds are pushed through the column. Because they spend *less time in the stationary phase,* they don't separate as well, and the GC *peaks come out very sharp but not well separated.* If you *slow the carrier gas down too much,* the compounds spend so much time in the stationary phase that *the peaks broaden and overlap gets very bad.* The optimum is, as always, the best separation you need, in the shortest amount of time. Sometimes you're on your own. Most of the time, someone else has already worked it out for you.

Temperature

Whether or not you realize it, the GC column has its own heater—the **column oven**. If you turn the temperature up, the compounds hotfoot it through the column very quickly. Because they spend *less time in the stationary phase,* they don't separate as well, and the GC *peaks come out very sharp but not well separated.* If you *turn the temperature down some,* the compounds spend so much time in the stationary phase that *the peaks broaden and overlap gets very bad.* The optimum is, as always, the best separation you need in the shortest amount of time. There are two absolute limits, however.

1. *Too high a temperature, and you destroy the column.* The adsorbent may decompose, or the liquid phase may boil out onto the detector. *Never exceed the recommended maximum temperature for the column material.* Don't even come within 20°C of it.
2. *Way too low a temperature, and the material condenses on the column.* You have to be above the **dew point** of the least volatile material. Not the boiling point. Water doesn't always condense on the grass—

become dew—every day that's just below 100°C (that's 212°F, the boiling point of water). Fortunately, you don't have to know the dew points for your compounds. However, you *do* have to know that you *don't have to be above the boiling point* of your compounds

Incidentally, the **injector** may have a separate **injector oven**, and the **detector** may have a separate **detector oven.** Set them both 10 to 20°C higher than the column temperature. You can even set these *above the boiling points* of your compounds, since you *do not want them to condense in the injection port or the detector, ever.* For those with *only one temperature control,* sorry. The injection port, column, and detector are all in the same place, all in the same oven, and all at the same temperature. The *maximum temperature,* then, is *limited by the decomposition temperature of the column.* Fortunately, because of that dew point phenomenon, you really don't have to work at the boiling points of the compounds either.

30

HP LIQUID
CHROMATOGRAPHY

HPLC. Is it **high-performance liquid chromatography** or **high-pressure liquid chromatography,** or something else? It's probably easier to consider it a delicate blend of **wet-column chromatography** and **gas chromatography** (see Chapters 20 and 24, respectively).

Rather than letting gravity **pull** the solvent through the powdered adsorbent, the liquid is **pumped** through under pressure. Initially, high pressures [1000–5000 psig (**pounds per square inch gauge**), i.e., not absolute] were used to push a liquid through a tightly packed solid. But the technique works well at lower pressures (250 psig), hence the name high-performance liquid chromatography.

From there on, the setup (Fig. 141) resembles gas chromatography very closely. Although a **moving liquid phase** replaces the helium stream, compounds are put onto a column through an **injection port**. They *separate inside a chromatographic column* in the same way as in GC by spending more or less time in a *moving liquid* now, and the separated compounds pass *through a detector*. There the amounts of each compound as they go by the **detector** are turned into electrical signals and *displayed on a chart recorder as HPLC peaks* that look just like GC peaks. You should also get the feeling that the analysis of these HPLC traces is done in the same manner as GC traces, because it is.

Again, there's a lot of variety among HPLC systems, so what I say won't necessarily apply to your own system in every respect. But it should help. I've based my observations on a Glencoe HPLC unit. It is simple and rugged, performs very well, and uses very common components carried by almost every HPLC supplier. Parts are easy to get.

That's probably why the company has stopped making this unit. Anything simple and rugged is not likely to need a lot of attention, nor is it likely to go out of fashion, and there's little profit in that. The individual pieces can be bought from many chromatographic supply houses. Just follow the directions for connections given here.

THE MOBILE PHASE: LIQUID

If you use only *one liquid*, either **neat** or as a mixture, the entire chromatogram is said to be **isochratic**. There are units that can deliver *varying solvent compositions over time*. These are called **gradient elution systems**.

For an **isochratic** system, you usually use a single solution, or a neat liquid, and put it into the **solvent reservoir**, which is generally a glass

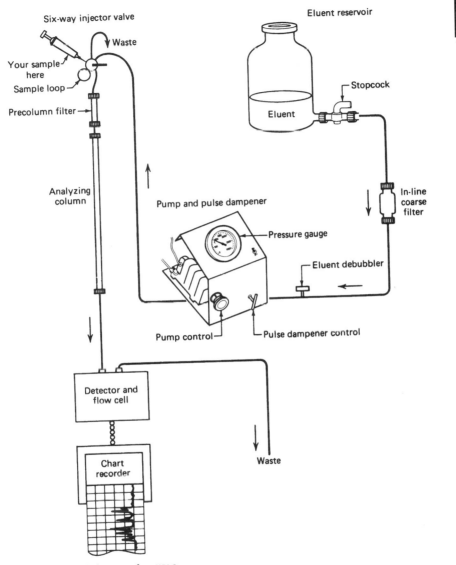

Fig. 141 Block diagram of an HPLC setup.

bottle with a stopcock at the bottom to let the solvent out (Fig. 142). The solvent travels out of the bottom of the reservoir and usually through a **solvent filter** that traps out any fairly large, insoluble impurities that may be in the solvent.

It is important, if you're making up the eluent yourself, to follow the directions *scrupulously*. Think about it. If you wet the entire system with

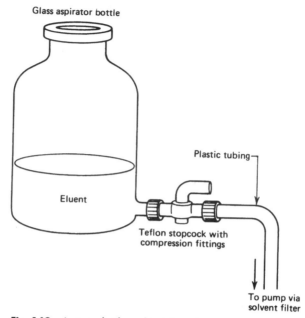

Fig. 142 Aspirator bottle used to deliver eluent.

the *wrong eluent*, you can wait a very long time for the *correct* eluent to reestablish the correct conditions.

A Bubble Trap

Air bubbles are the nasties in HPLC work. They cause the same type of troubles as they do in **wet-column chromatography**, and you just don't want them. So there's usually a **bubble trap** (Fig. 143) before the eluent reaches the pump. This device is really quite simple. Bubbles in the eluent

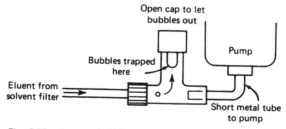

Fig. 143 Common bubble trap.

stream rise up the center pipe and are trapped there. To get rid of the bubbles, you open the cap at the top. Solvent then rises in the tube and pushes the bubbles out. You have to be *extremely* careful about bubbles if you're the one to start the setup or if the solvent tank has run out. Normally, *one bubble purge per day* is enough.

The Pump

The most common pumping system is the **reciprocating pump**. Milton Roy makes a pretty good model. The pump has a **reciprocating ruby rod** that moves back and forth. On the backstroke, the pump loads up on a little bit of solvent; then it squirts it out, under pressure, on the forward stroke. If you want to increase the amount of liquid going through the system, you can dial the length of the stroke, from zero to a preset maximum, using the micrometer at the front of the pump (see Fig. 144). Use a *fully clockwise* setting, and the stroke length is zero—*no solvent flow. A fully counterclockwise* setting gives the maximum stroke length—*maximum solvent flow*. If you have a chance to work with this type of pump, *always turn the micrometer fully clockwise to give a zero stroke length before you start the pump*. If there *is* a stroke length set *before* you turn the pump on, the first smack can damage the reciprocating ruby rod. And it is *not* cheap.

The Pulse Dampener

Because the rod **reciprocates** (i.e., goes back and forth), you'd expect huge swings in pressure, and pulses of pressure to occur. That's why they make **pulse dampeners**. A coiled tube is hooked to the pump on the side opposite the column (Fig. 144). It is filled with the eluent that's going through the system. On the forward stroke, solvent is compressed into this tube, and *at the same time*, a shot of solvent is pushed onto the column. On the backstroke, while the pump chamber fills up again, the eluent we just pressed into the pulse dampener *squirts out into the column*. Valves in the pump take care of directing the flow. With the eluent in the pulse dampener tubing taking up the slack, the huge variations in pressure, from essentially zero to perhaps 1000 psig, are evened out. They don't disappear, going about 100 psig either way, but these **dampened pulses** are now too small to be picked up on the detector. They don't show up on the chart recorder either.

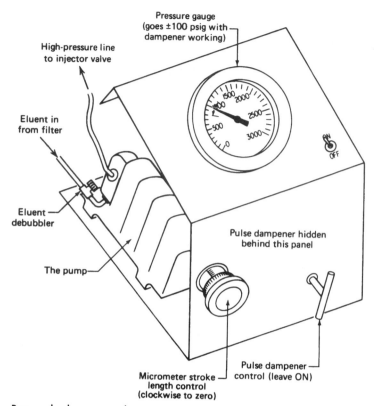

High-pressure line
to injector valve

Pressure gauge
(goes ±100 psig with
dampener working)

Eluent in
from filter

Eluent
debubbler

Pulse dampener hidden
behind this panel

The pump

Micrometer stroke
length control
(clockwise to zero)

Pulse dampener
control (leave ON)

Fig. 144 Pump, pulse dampener, and pressure-gauge unit.

HPLC SAMPLE PREPARATION

Samples for HPLC must be *liquids* or *solutions*. It would be nice if the solvent in which you've dissolved your solid sample were the same as the eluent.

It is *absolutely crucial* that you preclean your sample. *Any* decomposed or insoluble material will stick to the top of the column and can continually poison further runs. There are a few ways to keep your column clean.

1. ***The Swinney adapter*** (Fig. 145). This handy unit locks onto a syringe *already* filled with your sample. Then you push the sample *slowly* through a Millipore filter to trap insoluble particles. This does *not*, however, get rid of **soluble tars** that can ruin the column. (Oh. Don't confuse the filters with the papers that separate them. It's embarrassing.)

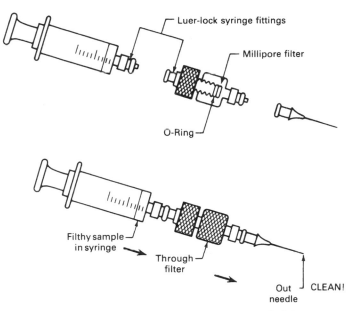

Fig. 145 The Swinney adapter and syringe parts. (Note Luer Lock.)

2. *The precolumn filter* (Fig. 146). Add a tiny column, filled with exactly the same material as the main column, and *let this small column get contaminated*. Then unscrew it, clean out the gunk and adsorbent, and refill it with fresh column packing. The disadvantage is that you don't really know when the garbage is going to poison the entire precolumn filter and then start ruining the analyzing column. The only way to find out whether you have to clean the precolumn is to take it out of the instrument. You really want to clean it out long before the contaminants start to show up at the precolumn exit.

HPLC SAMPLE INTRODUCTION

This is the equivalent of the **injection port** for the GC technique. With GC you could inject through a rubber septum directly onto the column. With HPLC it's very difficult to inject against a liquid stream moving at possibly 1000 psig. That's why **injection port valves** were invented for HPLC: you put your sample into an **injection loop** on the valve that is *not in the liquid stream*, then *turn the valve*, and *voilà, your sample is in the stream*, headed for the column.

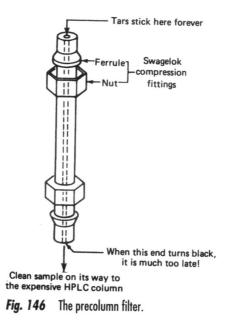

Tars stick here forever

Ferrule
Nut
Swagelok
compression
fittings

When this end turns black,
it is much too late!

Clean sample on its way to
the expensive HPLC column

Fig. 146   The precolumn filter.

The valve (Fig. 147) has two positions.

1. ***Normal solvent flow.*** In this position, the eluent comes into the valve, goes around, and comes on out into the column without any bother. *You put the sample on the column in this position.*
2. ***Sample introduction.*** Flipped this way, the eluent is *pumped through the sample loop* and any sample there is carried along and into the column. *You put the sample on the column in this position.*

SAMPLE IN THE COLUMN

Once the sample is in the column, there's not much difference between what happens here and what happens in paper, thin-layer, vapor-phase (gas), wet-column, or dry-column chromatography. *The components in the mixture will remain in the stationary phase, or move in the mobile phase for different times and end up at different places when you stop the experiment.*

So what's the advantage? You can separate and detect microgram quantities of solid samples much as in GC. And you can't do solids all that well by GC because you have to vaporize your solid sample, probably decomposing it.

31

A novel development for HPLC is something called **bonded reversed-phase columns,** where the stationary phase is a nonpolar hydrocarbon, chemically bonded to a solid support. You can use these with aqueous eluents, usually alcohol–water mixtures. So you have a *polar eluent and a nonpolar stationary phase,* something that does not usually occur for ordinary wet-column chromatography. One advantage is that you don't need to use anhydrous eluents (very small amounts of water can change the character of normal-phase columns) with reversed-phase columns.

SAMPLE AT THE DETECTOR

Many HPLC detectors can turn the presence of your compound into an electrical signal to be written on a chart recorder. At the time the **refractive index** detector was common. Clean eluent, *used as a reference,* went through one side of the detector, and the *eluent with the samples* went through the other side. *A difference in the refractive index between the sample and reference caused an electrical signal to be generated and sent to a chart recorder.* If you've read the section on **gas chromatography** and looked ahead at **infrared,** you shouldn't be surprised to find both a **sample** and a **reference.** I did tell you the reference/sample pair is common in instrumentation.

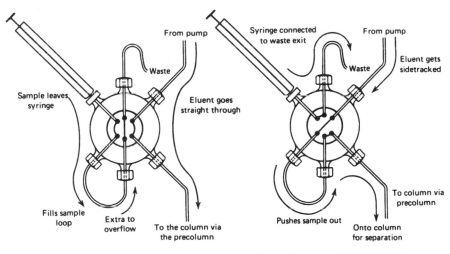

Fig. 147 The sample injector valve.

More recently, **UV detectors** (Fig. 148) have become more popular. UV radiation is beyond the purple end of the rainbow, the energy from which great tans are made. So, if you set up a **small mercury vapor lamp** with a power supply to light it up, you'll have a **source of UV light**. It's usually filtered to let through only the wavelengths of 180 and/or 254 nm. (And *where* have you met *that* number before? TLC plates, maybe?) This UV light then passes through a **flow cell** that has the *eluent and your separated sample flowing through it* against air as a reference. *When your samples come through the cell, they absorb the UV, and an electrical signal is generated.* Yes, the signal goes to a chart recorder and shows up as HPLC peaks.

SAMPLE ON THE CHART RECORDER

Go back and read about HPLC peak interpretation in the section on GC peak interpretation ("Sample on the Chart Recorder" in Chapter 30). The analysis is *exactly the same*, retention times, peak areas, baselines, ... all that.

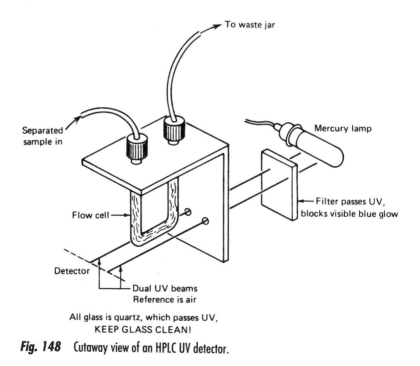

Fig. 148 Cutaway view of an HPLC UV detector.

PARAMETERS, PARAMETERS

31

To get the best LC trace from a given column, there are lots of things you can do; most of them are the same as those for GC (see "Gas Chromatography, Parameters, Parameters" in Chapter 30).

Eluent Flow Rate

The faster the eluent flows, the faster the compounds are pushed through the column. Because they spend *less time in the stationary phase*, they don't separate as well and the *LC peaks come out very sharp but not well separated.* If you *slow the eluent down too much,* the compounds spend so much time in the stationary phase that the *peaks broaden and overlap gets very bad.* The optimum is, as always, the best separation you need, in the shortest amount of time. One big difference in LC is the need to worry about **back-pressure**. If you try for very high flow rates, the LC column packing tends to collapse under the pressure of the liquid. This, then, is the cause of the back-pressure: *resistance of the column packing.* If the pressures get *too high*, you may burst the tubing in the system, damaging the pump, ... all sorts of fun things.

Temperature

Not many LC setups have ovens for temperatures like those for GC. This is because *eluents tend to boil at temperatures much lower than the compounds on the column,* which are usually solids anyway. Eluent bubbling problems are bad enough, without actually boiling the solvent in the column. This is not to say that LC results are independent of temperature. They're not. But if a *column oven for LC is present*, its purpose more likely is *to keep stray drafts and sudden chills away* than to have a hot time.

Eluent Composition

You can vary the composition of the eluent (mobile phase) a lot more in HPLC than in GC, so there's not really much correspondence. Substitute nitrogen for helium in GC and usually the sensitivity decreases, but the retention times stay the same. Changing the mobile phases—the gases—in GC doesn't have a very big effect on the separation or retention time.

There are much better parallels to HPLC: **TLC** or **column chromatography**. Vary the eluents in these techniques, and you get widely different results. With a **normal-phase silica-based column**, you can get results similar to those from silica gel TLC plates.

INFRARED
SPECTROSCOPY

Unlike the chromatographies, which physically separate materials, **infrared (IR) spectroscopy** is a method of determining what you have after you've separated it.

The **IR spectrum** is the name given to a band of frequencies between 4000 and 650 cm^{-1} beyond the red end of the visible spectrum. The units are called **wave numbers** or **reciprocal centimeters** (that's what cm^{-1} means). This range is also expressed as wavelengths from 2.5 to 15 micrometers (μ).

With your sample in the **sample beam**, the instrument scans the IR spectrum. *Specific functional groups absorb specific energies.* Because the spectrum is laid out on a piece of paper, these specific energies become *specific places* on the chart.

Look at Fig. 149. Here's a fine example of a pair of alcohols if ever there was one. See the peak (some might call it a trough) at about 3400 cm^{-1}(2.9 μ)? That's due to the **OH group**, specifically the stretch in the O—H bond, the **OH stretch.**

Now consider a couple of ketones, 2-butanone and cyclohexanone (Fig. 150). There's no OH peak about 3400 cm^{-1}(2.9μ), is there? Should there be? *Of course not.* Is there an OH in 2-butanone? *Of course not.* But there is a C=O, and where's that? The peak about 1700 cm^{-1}(5.9μ). It's *not there* for the alcohols, and it *is* there for the ketones. Right. You've just correlated or *interpreted* four IRs.

Because the first two (Fig. 149) have the *characteristic OH stretch of alcohols*, they might just be alcohols. The other two (Fig. 150) might be ketones because of the *characteristic C=O* stretch at 1700 cm^{-1}(5.9μ) in each.

What about all the other peaks? You *can* ascribe some sort of meaning to each of them, but it can be very difficult. That's why **frequency correlation diagrams,** or **IR tables**, exist (Fig. 151). They identify regions of the IR spectrum where peaks for various functional groups show up. They can get very complicated. Check to see if you can find the C—H stretch and the C—O stretch that are in all four spectra, using the correlation table. It can be fun.

For you Sherlock Holmes fans, the region from 1400 to 990 cm^{-1}(7.2 to 11.1 μ) is known as the **fingerprint region**. The peaks are due to the entire molecule, *its fingerprint,* rather than being from independent functional groups. And, you guessed it, no two fingerprints are alike.

Take another look at the cyclohexanol and cyclohexanone spectra. Both have very different functional groups. Now look at the similarities, the simplicity, including the fingerprint region. Both are six-membered rings

32

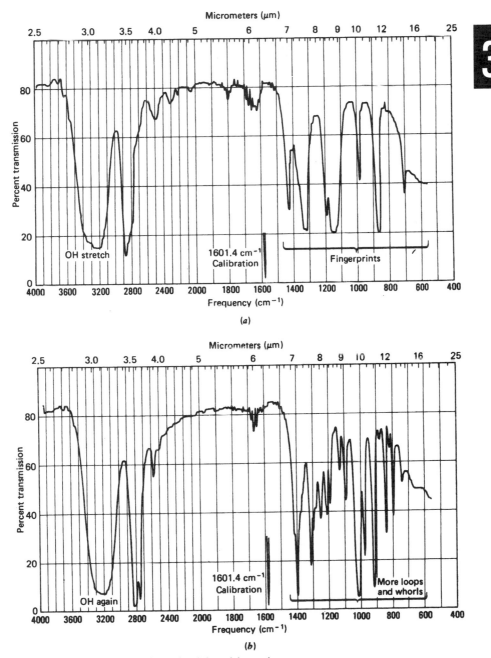

Fig. 149 IR spectra of a (a) t-butanol and (b) cyclohexanol.

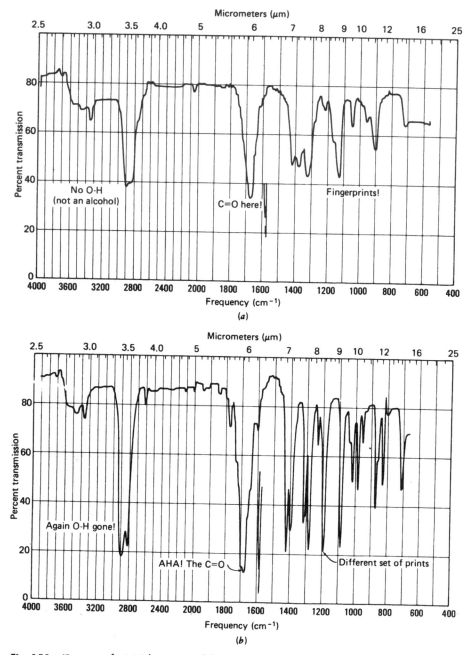

Fig. 150 IR spectra of a (a) 2-butanone and (b) cyclohexanone.

32

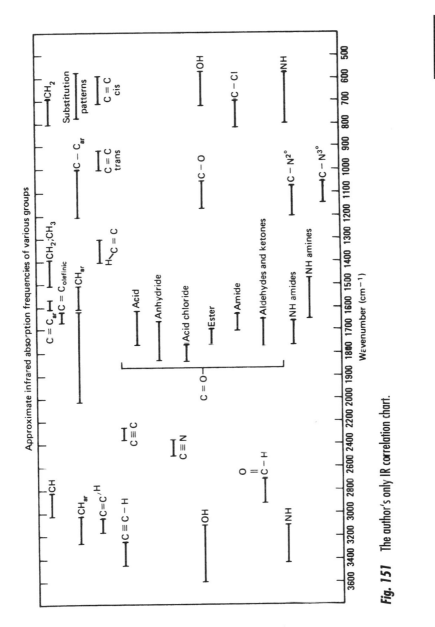

Approximate infrared absorption frequencies of various groups

Fig. 151 The author's only IR correlation chart.

and have a high degree of symmetry. You should be able to see the similarities owing to the similar structural features.

Two more things. First, watch your spelling and pronunciation; it's not "infared," OK? Second, most people I know use IR (pronounced "eye-are," not "ear") to refer to the technique, the instrument, and the chart recording of the spectrum:

"That's a nice new IR you have there." (the instrument)
"Take an IR of your sample." (perform the technique)
"Let's look at your IR and see (interpret the resulting
    what kind of compound you have." spectrogram)

To take an IR, you need an IR. These are fairly expensive instruments; again, no one is typical, but you can get a feeling of how to run an IR as you go on.

INFRARED SAMPLE PREPARATION

You can prepare samples for IR spectroscopy easily, but you must strictly adhere to one rule:

No water!

In case you didn't get that the first time:

No water!

Ordinarily, you put the sample between two salt plates. Yes. Common, ordinary *water-soluble* salt plates. Or mix it with **potassium bromide (KBr),** another water-soluble salt.

So keep it dry, people.

Liquid Samples

1. Make sure the sample is DRY. NO WATER!
2. Put some of the *dry* sample (2–3 drops) on one plate, then cover it with another plate (Fig. 152). The sample should spread out to cover the entire plate. You *don't have to press*. If it doesn't cover well, try turning the top plate to spread the sample, or add more sample.
3. Place the sandwich in the IR salt plate holder and cover it with a hold-down plate.

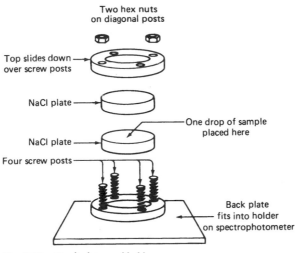

DO NOT OVERTIGHTEN

Two hex nuts
on diagonal posts

Top slides down
over screw posts

NaCl plate

One drop of sample
placed here

NaCl plate

Four screw posts

Back plate
fits into holder
on spectrophotometer

32

Fig. 152 IR salt plates and holder.

4. Put at least two nuts on the posts of the holder (opposite corners) and spin them down *GENTLY* to hold the plates with an even pressure. *Do not use force!* You'll crack the plates! Remember, these are called salt plate holders and not salt plate smashers.
5. Slide the holder and plate into the bracket on the instrument in the sample beam (closer to you, facing the instrument.)
6. Run the spectrum.

Since you don't have any other solvents in there, just your liquid compound, you have just prepared a liquid sample, **neat**, meaning *no solvent*. This is the same as a *neat liquid sample,* which is a way of describing any liquid without a solvent in it. It is not to be confused with a "really neat liquid sample," which is a way of expressing your true feelings about your sample.

Solid Samples

The Nujol Mull

A rapid, inexpensive way to get an IR of solids is to mix them with Nujol, a commercially available mineral oil. Traditionally, this is called "making a Nujol mull," and it is practically idiomatic among chemists. Although you won't see Rexall or Johnson & Johnson mulls, the generic brand **mineral oil mull** is often used.

You want to disperse the solid throughout the oil, making the solid transparent enough to IR that the sample will give a usable spectrum. Since mineral oil is a saturated hydrocarbon, it has an IR spectrum all its own. You'll find hydrocarbon bends, stretches, and pushups in the spectrum, but you know where they are, and you ignore them. You can either look at a published reference Nujol spectrum (Fig. 153) or run your own if you're not sure where to look.

1. Put a small amount of your solid into a tiny agate mortar and add a few drops of mineral oil.
2. Grind the oil and sample together until the solid is a fine powder *dispersed throughout the oil.*
3. Spread the mull on one salt plate and cover it with another plate. There should be no air bubbles, just an even film of the solid in the oil.
4. Proceed as if this were a *liquid sample.*
5. Clean the plates with *anhydrous* acetone or ethanol. *NOT WATER!* If you don't have the tiny agate mortar and pestle, try a Witt spot plate and the rounded end of a thick glass rod. The spot plate is a piece of glazed porcelain with dimples in it. Use one as a tiny mortar and the other as a tiny pestle.

And remember to forget the peaks from the Nujol itself.

Solid KBr Methods

KBr methods (hardly ever called potassium bromide methods) consist of making a mixture of your solid *(dry again)* with IR-quality KBr. Regular KBr off the shelf is likely to contain enough nitrate, as KNO_3, to give spurious peaks, so don't use it. After you have opened a container of KBr, dry it and later store it in an oven, with the cap off, at about 110°C to keep the moisture out.

Preparing the Solid Solution

1. At least once in your life, *weigh out* 100 mg of KBr so you'll know how much that is. If you can remember what 100 mg of KBr looks like, you won't have to weigh it out every time you need it for IR.

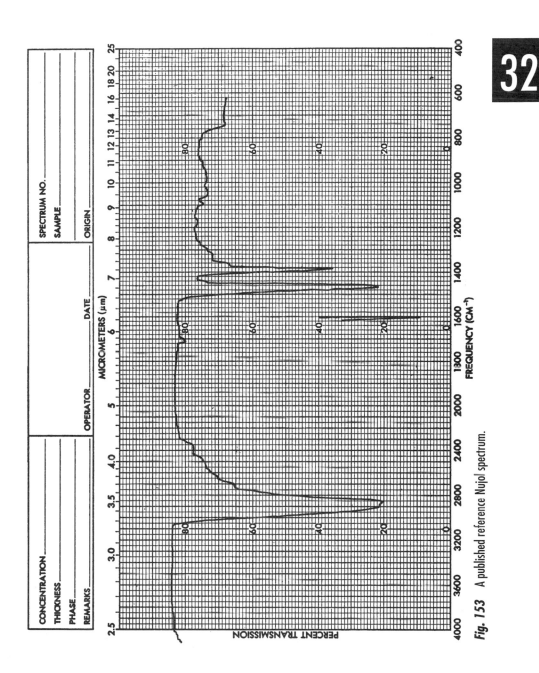

Fig. 153 A published reference Nujol spectrum.

32

2. Weigh out 1–2 mg of your *dry, solid* sample. You'll have to weigh out each *sample* because different compounds take up different amounts of space.

3. Pregrind the KBr to a fine powder about the consistency of powdered sugar. Don't take forever, since moisture from the air will be coming in.

Pressing a KBr Disk—The Mini-Press (Fig. 154)

1. Get a clean, dry press and two bolts. Screw one of the bolts into the press about halfway and call that the bottom of the press.

2. Scrape a finely ground mixture of your compound (1–2 mg) and KBr (approximately 100 mg) into the press so that an even layer covers the bottom bolt.

3. Take the other bolt and turn it in from the top. *Gently* tighten and loosen this bolt at least once to spread the powder evenly on the face of the bottom bolt.

4. Hand tighten the press again, and then use wrenches to tighten the bolts against each other. Don't use so much force that you turn the heads right off the bolts.

5. Remove both bolts. A KBr pellet, containing your sample, should be in the press. Transparent is excellent. Translucent will work. If the sample is opaque, you can run the IR, but I don't have much hope of your finding anything.

6. Put the entire press in a holder placed in the analyzing beam of the IR, as in Fig. 155. (Don't worry about that yet; I'll get to it in a moment. See "Running the Spectrum" in the next major section.)

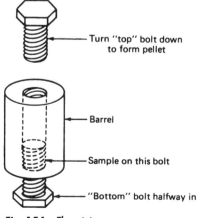

Turn "top" bolt down to form pellet

Barrel

Sample on this bolt

"Bottom" bolt halfway in

Fig. 154 The mini-press.

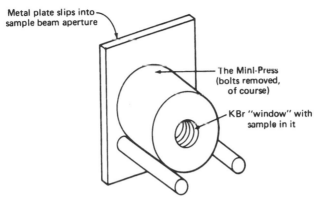

Fig. 155 The mini-press in its holder.

Pressing a KBr Disk—The Hydraulic Press

If you have a hydraulic press and two steel blocks available, there is another easy KBr method. It's a card trick, and at no time do my fingers leave my hands. The only real trick is you'll have to bring the card.

1. Cut and trim an index card so that it fits into the **sample beam aperture** (Fig. 160).
2. Punch a hole in the card with a paper punch. The hole should be *centered in the sample beam* when the card is in the sample beam aperture.
3. Place one of the two steel blocks on the bottom jaw of the press.
4. Put the card on the block.
5. Scrape your KBr-sample mixture onto the card, covering the hole and some of the card. Spread it out evenly.
6. Cover the card, which now has your sample on it, with the second metal block (Fig. 156).
7. Pump up the press to the indicated *safe* pressure.
8. Let this sit for a bit (1 minute). If the pressure has dropped, bring it back up, *slowly, carefully,* to the safe pressure line, and wait another bit (1 minute).
9. Release the pressure on the press. Push the jaws apart.
10. Open the metal sandwich. Inside will be a file card with a **KBr window** in it, just like in the Mini-Press.

CAUTION! *The KBr window you form is rather fragile, so don't beat on it.*

11. Put the card in the **sample beam aperture** (Fig. 157). The KBr window should be *centered* in the sample beam if you've cut and punched the card correctly.

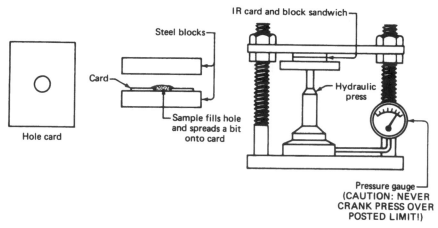

Fig. 156 KBr disks by hydraulic press.

RUNNING THE SPECTRUM

There are many IR instruments, and since they are so different, you need your instructor's help here more than ever. But there are a few things you have to know.

1. *The sample beam*. Most IRs are **dual-beam instruments** (Fig. 158). The one closest to you, if you're operating the instrument, is the **sample beam.** Logically, there is a **sample holder** for the sample beam, and your sample goes there. And there, a beam of IR radiation goes through your sample.

2. *The reference beam*. This is the other light path. It's not *visible* light but another part of the electromagnetic spectrum. Just remember that the reference beam is the one farthest away from you.

3. *The 100% control*. This sets the pen at the 100% line on the chart paper. Or tries to. It's a very delicate control and doesn't take kindly to excessive force. Read on, and all will be made clear.

4. *The pen.* There is a pen and pen holder assembly on the instrument. This is how the spectrum gets recorded on the chart paper. Many people get the urge to throw the instrument out the window when the pen stops writing in the middle of the spectrum. Or doesn't even start writing. Or was left empty by the last fellow. Or was left to dry out on the top of the instrument. For those with Perkin-Elmer 137s and 710s, two clever fellows have made up generic felt-tip pen holders for the instruments. This way, you buy your own pen from the bookstore, and

32

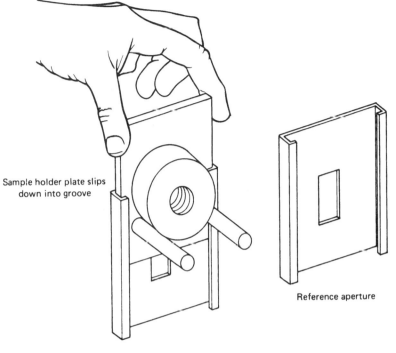

Sample holder plate slips
down into groove

Reference aperture

Sample aperture

Fig. 157 Putting sample holders with samples into the beam.

if it dries out it's your own fault. [See R. A. Bailey and J. W. Zubrick, *J. Chem. Educ.*, 59, 21 (1982).]

5. ***The very fast or manual scan.*** To get a good IR, you'll have to be able to scan, very rapidly, *without letting the pen write on the paper*. This is so you'll be able to make adjustments before you commit pen to paper. This **fast forward** is not a standard thing. Sometimes you operate the instrument by hand, pushing or rotating the paper holder. Again, *do not use force*.

Whatever you do, *don't try to move the paper carrier by hand when the instrument is scanning a spectrum*. Stripped gears is a crude approximation of what happens. So, thumbs off.

I'm going to apply these things in the next section using a real instrument, the Perkin-Elmer 710B, as a model. Just because you have another model is no reason to skip this section. If you do have a different IR, try to find the similarities between it and the Perkin-Elmer model. Ask your instructor to explain any differences.

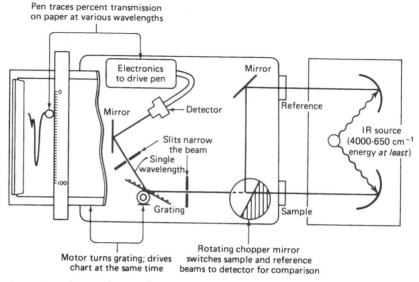

Fig. 158   Schematic diagram of an IR.

THE PERKIN-ELMER 710B IR (FIG. 159)

1. ***On-off switch and indicator.*** Press this once, the instrument comes on, and the switch lights up. Press this again, the instrument goes off, and the light goes on.

2. ***Speed selector.*** Selects speed (normal or fast). "Fast" is faster, but slower gives *higher resolution*, that is, more detail.

3. ***Scan control.*** Press this to *start a scan*. When the instrument is scanning, the optics and paper carrier move automatically, causing the IR spectrum to be drawn on the *chart paper* by the *pen*.

4. ***Chart paper carriage.*** This is where the chart paper nestles while you run an IR. If it looks suspiciously like a clipboard, it's because that's how it works.

5. ***Chart paper hold-down clip.*** Just like a clipboard, this holds the paper down in the carrier.

6. ***Frequency scale.*** This scale is used to help align the chart paper and to tell you during the run where in the spectrum the instrument is.

7. ***Scan position indicator.*** A white arrow that points, roughly, to where the instrument is in the spectrum.

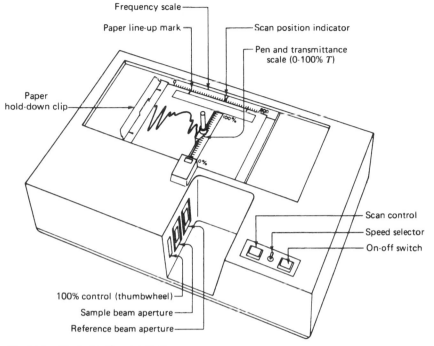

Fig. 159 The Perkin-Elmer 710B IR.

8. *Line-up mark.* A line, here at the number 4000, that you use to match up the numbers on the instrument frequency scale with the same numbers on the chart paper.

9. *Pen and transmittance scale.* This is where the pen traces your IR spectrum. The numbers here mean percentage of IR transmitted through your sample. If you have no sample in the sample beam, how much of the light is getting through? Those who said 100% are 100% correct. Block the beam with your hand and 0% gets through. You should be able to see why these figures are called the % **transmission**, or %**T**, scale.

10. *The 100% control.* Sets the pen at 100%. Or tries to. This is a fairly sensitive control, so don't force it.

11. *Sample beam aperture.* This is where you put the holder containing your sample, be it mull or KBr pellet. You slip the holder into the aperture window for analysis.

12. *Reference beam aperture.* This is where nothing goes. Or, in extreme cases, you use a **reference beam attenuator** to cut down the amount of light reaching the detector.

USING THE PERKIN-ELMER 710B

1. Turn the instrument on and let it warm up for about 3–5 minutes. Other instruments may take longer.
2. Get a piece of IR paper and load the *chart paper carriage*, just like a clipboard. Move the paper to get the index line on the paper to line up with the index line on the instrument. It's at 4000cm^{-1} and it's only a rough guide. Later I'll tell you how to calibrate your chart paper.
3. Make sure the chart paper carriage is at the high end of the spectrum 4000cm^{-1}.
4. Put your sample in the sample beam. Slide the sample holder with your sample into the sample beam aperture (Fig. 157).
5. What to do next varies for particular cases—not much but enough to be confusing in setting things up.
6. Look at where the pen is. *Carefully* use the **100% control** to locate the pen at about the 90% mark when the chart (and spectrum) is at the high end 4000cm^{-1}.

The 100% Control: An Important Aside

Usually, there's not much more to adjusting the 100% control than is shown in upcoming items 7 through 9, unless your sample, *by its size alone,* reduces the amount of IR reaching the detector. This really shows up if you've used the mini-press, which has a *much smaller opening* than that of the opening in the reference beam. So you're at a disadvantage right from the start. Sometimes the 100% control mopes around at much less than 40%. That's terrible, and you'll have to use a **reference beam attenuator** (Fig. 160) to equalize the amounts of energy in the two beams. As you block more and more of the energy in the reference beam, the %T will go back to the 100% mark. Stop at about 90%. Note that, if you didn't have these problems, this is where you'd put the pen with the 100% control anyway. Use the *smallest amount* of reference beam attenuation you can get away with.

7. OK, at 4000cm^{-1} the %T (the pen) is at 90%.
8. Now *slowly, carefully,* move the pen carriage manually so that the instrument scans the entire spectrum. *Watch the pen!* If the **baseline**

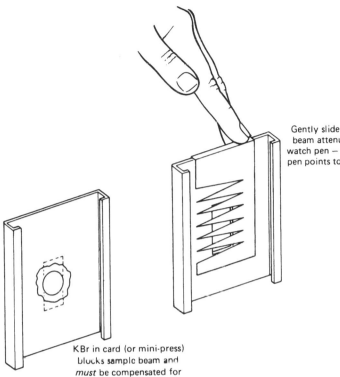

Gently slide reference beam attenuator and watch pen — stop when pen points to 80-90% *T*

KBr in card (or mini-press) blocks sample beam and *must* be compensated for

Fig. 160 Using a reference beam attenuator with a KBr window.

creeps up and goes off the paper (Fig. 161), this is not good. Readjust the 100% control to keep the pen on the paper. Now keep going, slowly, and *if* the pen drifts up again, readjust it again with the 100% control and get the pen back on the paper.

9. Now go back to the beginning (4000 cm⁻¹). If you've adjusted the 100% control to get the pen back on the paper at some other part of the spectrum, surprise! The pen will not be at 90% when you get back. This is unimportant. What is important is that *the pen stay within the limits,* between the *0* and *100%T* lines, *for the entire spectrum.*

10. In *any* case, if the peaks are too large, with the baseline in the proper place, your sample is just *too concentrated.* You can wipe some of your liquid sample or mull off one of the salt plates or remake the KBr pellet using less compound or more KBr. Sorry.

11. When you've made all the adjustments, press the scan button and you're off.

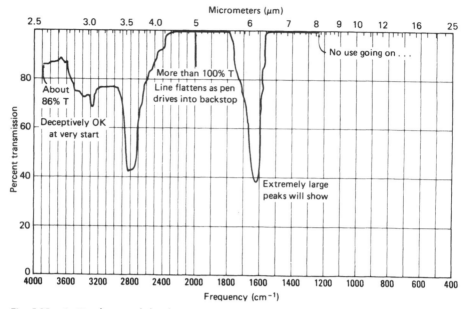

Fig. 161 An IR with an unruly baseline.

CALIBRATION OF THE SPECTRUM

Once the run is over, there's one other thing to do. Remember that the index mark on the paper is not exact. You have to **calibrate the paper** with a standard, usually *polystyrene film.* Some of the peaks in polystyrene are quite sharp, and many of them are very well characterized. A popular one is an extremely narrow, very sharp spike at 1601.4 cm⁻¹ (6.24μ).

1. *Don't move the chart paper,* or this calibration will be worthless.
2. Remove your sample, and replace it with the standard polystyrene film sample. You will have to remove any reference beam attenuator and turn the 100% control to set the pen at about 90%, when the chart is at 4000 cm⁻¹.
3. *With the pen off the paper*, move the carriage so that it's *just before* the calibration peak you want, in this case 6.24 μ.
4. Now, quickly, start the scan, *let the pen draw just the tip of the calibration peak,* and quickly stop the scan. You don't need to draw more than that (Fig. 162). Just make sure you can pick your calibration peak out of the spectral peaks. If it's too crowded at 1601.4 cm⁻¹, use a different polystyrene peak—2850.7 or 1028.0 cm⁻¹

(3.51 or 9.73 μ). Anything really well known and fairly sharp will do (Fig. 163).

5. And that's it. You have a nice spectrum.

32

IR SPECTRA: THE FINISHING TOUCHES (FIG. 164)

IR chart paper contains spaces for all sorts of information. It would be nice if you could fill in

1. ***Operator.*** The person who ran the spectrum. Usually you.
2. ***Sample***. The name of the compound you've just run.
3. ***Date.*** The day you ran the sample.
4. ***Phase.*** For KBr, say "solid KBr." A Nujol mull is "Nujol mull." Liquids are either solutions in solvents or "neat liquids," that is, without any solvents, so call them liquids.
5. ***Concentration.*** For KBr, a solid solution, list milligrams of sample in 100 mg of KBr. For liquids, *neat* is used for liquids without solvents.

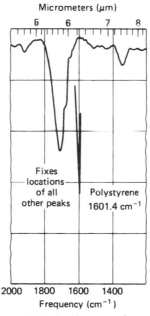

Fig. 162 A calibration peak on an IR spectrum.

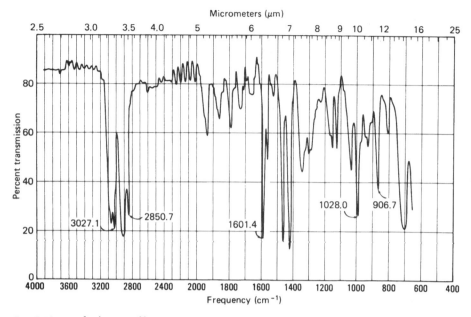

Fig. 163 IR of polystyrene film pointing out many calibration peaks.

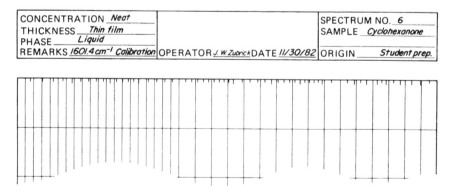

CONCENTRATION *Neat*		SPECTRUM NO. *6*
THICKNESS *Thin film*		SAMPLE *Cyclohexanone*
PHASE *Liquid*		
REMARKS *1601.4 cm⁻¹ Calibration*	OPERATOR *J. W. Zubrick* DATE *11/30/82*	ORIGIN *Student prep.*

Fig. 164 The finishing touches on the IR.

32

6. ***Thickness.*** Unless you're using solution cells, write **thin film** for **neat liquids.** Leave this blank for KBr samples (unless you've measured the thickness of the KBr pellet, which you shouldn't have done).

7. ***Remarks.*** Tell where you put your calibration peak, where the sample came from, and anything unusual that someone in another lab might have trouble with when trying to duplicate your work. Don't put this off until the last day of the semester when you can no longer remember the details. Keep a record of the spectrum *in your notebook.*

You now have a perfect IR, suitable for framing and interpreting.

INTERPRETING IR's

IR interpretation can be as simple or as complicated as you'd like to make it. You've already seen how to distinguish alcohols from ketones by **correlation** of the positions and intensities of various peaks in your spectrum with positions listed in **IR tables** or **correlation tables**. This is a fairly standard procedure and is probably covered very well in your textbook. The things that are not in your text are as follows.

1. ***Not forgetting the Nujol peaks.*** Mineral oil will give huge absorptions from all the C—H bonds. They'll be the biggest peaks in the spectrum. And every so often, people mistake one of these for something that belongs with the sample.

2. ***Nitpicking a spectrum.*** Don't try to interpret every wiggle. There is a lot of information in an IR, but sometimes it is confusing. Think about what you're trying to show and, then show it.

3. **Negative results,** Negative results can be very useful. No big peak between 1600 and 2000 cm⁻¹? No carbonyl. Period. No exceptions.

4. ***Pigheadedness in interpretation.*** Usually a case of, "I know what this peak is so don't confuse me with facts." Infrared is an extremely powerful technique, but there are limitations. You don't have to go hog wild over your IR, though. I know of someone who decided that a small peak was an N-H stretch, and the compound *had* to have nitrogen in it. The facts that the intensity and position of the peak were not quite right, and neither a chemical test nor solubility studies indicated nitrogen, didn't matter. Oh well.

THE FOURIER TRANSFORM INFRA-RED (FTIR)

There's even more of a difference in particulars between FTIRs than between dispersive IRs (Like the 710B mentioned earlier), because not only can the basic instrument vary, but it can be attached to any number of different computers of differing configurations running any number of different software programs doing the infra-red analysis. Phew. So I'll only hit the major differences.

The Optical System

The optical system is based upon the Michelson Interferometer (Fig. 165). Infra-red energy from the source goes through the sample in a single beam and hits a beamsplitter in the interferometer. Half of the light gets reflected to a stationary mirror, and the other half passes through the beamsplitter to a moving mirror. Both mirrors reflect the light back to the beamsplitter, where they recombine to form an interference pattern of constructive and destructive interferences known as **interferogram.** (Consider the pattern produced when sets of waves from two stones thrown into a pond interact: The peak of one and the peak of the other re-inforce each other, giving a bigger peak; the peak of one and the trough of the other can completely destroy each other and produce no displacement.) That interference pat-

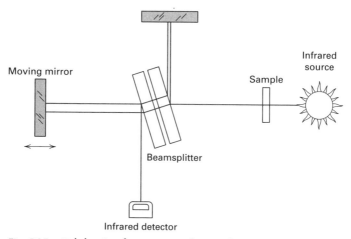

Fig. 165 Michelson Interferometer Optical System for FTIR.

tern varies with the displacement of the moving mirror, and this pattern of variation is sent along to the detector.

A computer programmed with the algorithm of the Fourier transformation converts the measured-intensity-versus-mirror-displacement signal (the interferogram) into a plot of intensity versus frequency—our friendly infra-red spectrum.

FTIR has a few advantages:

1. *Fellgett's Advantage (Multiplex Advantage)*. The interferometer doesn't separate light into individual frequencies like a prism or grating, so every point in the interferogram (spectrum) contains information from each wavelength of infra-red from the source.
2. *Jacquinot's Advantage (Throughput Advantage)*. The dispersive instrument, with prism or grating, needs slits and other optics so as much of a single wavelength of energy reaches the detector as possible. Not so the interferometer. So the entire energy of the source comes roaring through the optical system of the interferometer to the detector, increasing the signal-to-noise ratio of the spectrum.
3. *Conne's Advantage (Frequency Precision)*. The dispersive instrument depends on calibration (polystyrene at 1601 cm^{-1}), and the ability of gears and levers to move slits and gratings in a reproducable fashion. The FTIR carries its own internal frequency standard, usually a He–Ne gas laser, that serves as the master timing clock, tracking mirror movement and frequency calibration to a precision and accuracy of better than 0.01 wavenumbers (cm^{-1}).

But what about your advantages. And your disadvantages?

1. *FTIR is a single-beam instrument*. So you must collect a background spectrum (Figure 166) before you do your sample. The computer program will subtract the background from the background-and-sample and produce the usual IR spectrum. What's in a background spectrum? Everything that's not the sample. For a KBr disk spectrum, the empty press (Figure 155) or an empty card (Figure 156) gets included in the background spectrum. For a salt plate spectrum (Figure 152), clean salt plates in the holder is the background. And for a solution-cell spectrum, the cell with pure solvent is the background. Why? For many cells and holder combinations, the aperture is smaller, or the salt plates absorb a bit, or the solvent absorbs a lot, and these must be compensated for.

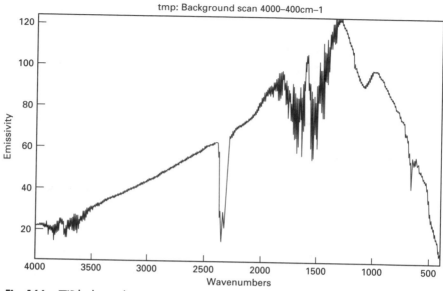

Fig. 166 FTIR background spectrum. Note H_2O and CO_2 bands.

2. **The FTIR is fast**. I can get the spectrum of a compound on my computer screen in about 0.21 minutes; considerably faster than any disperse instrument. Unfortunately, our inexpensive plotter is a bit slow, and, as the software makes the plotter label the axes with numbers and text, it does take a bit more than that 0.21 minutes. Overall, though, it is really considerably faster.

3. **The FTIR is accurate**. So you don't have to run a reference polystyrene. Unfortunately, we plot the spectra on inexpensive copier paper and if you don't mark relevant peaks on the screen with the software, you may never get those really accurate frequencies to 0.01 cm⁻¹.

4. **The FTIR computer programs automatically scale the spectrum**. So you don't have to set the zero and 100%. Unfortunately, the scaling is mindless, and you can be fooled if you're not careful. Figure 167 could possibly be an organic acid. See the broad hydrogen bonded OH peak at about 3400 cm⁻¹ (The consequences of not marking the exact frequency with the computer program should be evident now.) But the C=O at 1750 cm⁻¹ isn't as strong as it should be and .... Fooled ya! Look at the transmittance scale on the left hand side. It only goes from 70 to 76%. This is a spectrum of some student-abused salt plates and these huge peaks and bands are actually tiny wiggles (Figure 168.). The computer program mindlessly took the maximum and minimum wiggling in the spectrum and decided to give it to you at full-screen height.

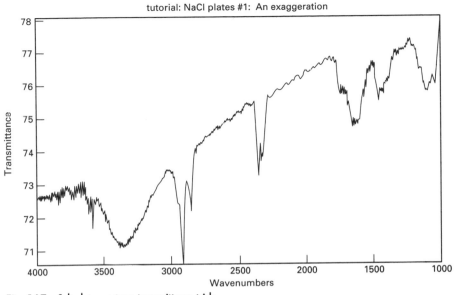

Fig. 167 Salt plate spectrum transmittance trick.

Admittedly, this is an unusual case. Since my background spectrum was taken with an empty salt-plate holder in the beam, my salt plates became the spectrum. Had I run the background with the holder and the cells, any bumps from the salt plates would have been cancelled.

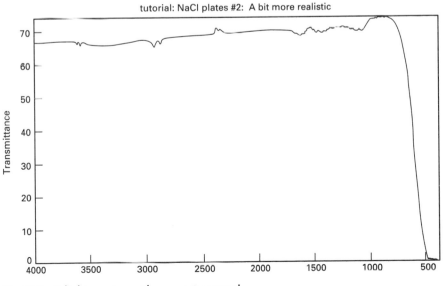

Fig. 168 Salt plate spectrum with some sanity restored.

And to get that 0 to 75% transmittance spectrum, I had to go outside the usual operating range of our instrument. Normally, you won't have such problems. But if you are now on your guard, you won't have *these* problems.

5. **The FTIR programs usually have a library spectral search, so you can easily identify your compounds.** Unfortunately, not every sample you'll run is in the database. So you could wind up with a collection of compounds usually having the same functional groups (and that can be helpful). Close—but no cigar. In that case, you must not succumb to the "I got it off the computer it must be right" syndrome.

NUCLEAR MAGNETIC RESONANCE (NMR)

Nuclear magnetic resonance (NMR) can be used like IR to help identify samples. But if you thought the instrumentation for IR was complicated, these NMR instruments are even worse. So I'll only give some generalities and directions for preparing samples.

For organic lab, traditionally you look only at the signals from protons in your compound; sometimes then, this technique is called **proton magnetic resonance (PMR)**. Not naked H^+ protons either, eh? The everyday **hydrogens** in organic compounds are just called **protons** when you use this technique.

A sample in a special tube is spun between the poles of a strong magnet. A radio-frequency signal, commonly 60 MHz, a little higher than TV Channel 2, is applied to the sample. Now, if *all* protons were in the same environment, there'd be this big absorption of energy in one place in the PMR spectrum. Big deal. But all protons are *not* the same. If they're closer to electronegative groups, or on aromatic rings, the signals shift to a different frequency. This change in the position of the PMR signals, which depends on the chemistry of the molecule, is called the **chemical shift**. Thus, you can tell quite a bit about a compound if you have its NMR.

LIQUID SAMPLE PREPARATION

To prepare a liquid sample for NMR analysis,

1. Get an **NMR tube**. These tubes are about 180 mm long and 5 mm wide, and the cost is about a buck apiece for what is euphemistically called the inexpensive model. The tubes are not precision ground, and some may stick in the NMR probe. This should not be your worry, though. They also have matching, color-coordinated designer caps (Fig. 169).

2. Get a **disposable pipet** and a little **rubber bulb** and construct a **narrow medicine dropper**. Use this to transfer your sample to the NMR tube. Don't fill it much higher than abut 3–4 cm. Without any solvent, this is called, of course, a **neat sample**.

3. Ask about an **internal standard**. Usually **tetramethysilane (TMS)** is chosen because most other proton signals from any sample you might have fall at *lower frequencies* than that of the protons in TMS. Sometimes **hexamethyldisiloxane (HMDS)** is used because it

33

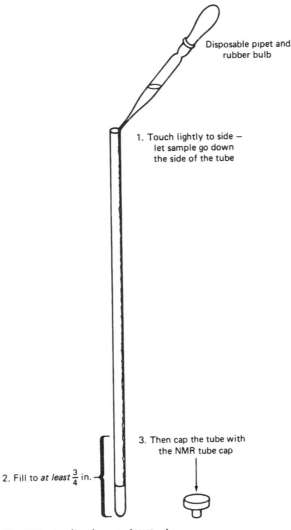

Disposable pipet and rubber bulb

1. Touch lightly to side — let sample go down the side of the tube

3. Then cap the tube with the NMR tube cap

2. Fill to *at least* $\frac{3}{4}$ in.

Fig. 169 Loading the typical NMR tube.

doesn't boil out of the NMR tube like TMS can. TMS boils at 26–28°C; HMDS boils at 101°C. Add *only* one or two drops.

4. Cap the tube and have the NMR of the sample taken. It's really out of place for me to tell you more about NMR here. Buy my next book, *If They Don't Work...They're Machines.*

5. Last point: cleanliness. If there is *trash* in the sample, get rid of it. Filter it or something, will you?

SOLID SAMPLES

Lucky you. You have a solid instead of a liquid. This presents one problem. What are you going to dissolve the solid in? Once it's a solution, you handle it just like a **liquid sample**. Unfortunately, if the *solvent has protons* and you know there'll be much more solvent than sample, you'll get a major proton signal from your solvent. Not a good thing, especially if the signals from your solvent and sample overlap.

Protonless Solvents

Carbon tetrachloride, a solvent without protons, is a typical protonless solvent. In fact, it's practically the only example. So if your sample dissolves in CCl_4, you're golden. Get at least 100 mg of your compound in enough solvent to fill the NMR tube to the proper height.

> **Caution!** CCl_4 is toxic and potentially carcinogenic. Handle with *extreme care.*

Deuterated Solvents

If your compound does not happen to dissolve in CCl_4, you still have a shot because *deuterium atoms do not give PMR signals*. This is logical, since they're not protons. The problem is that *deuterated solvents are expensive*, so do NOT ask for, say D_2O or $CDCl_3$, the **deuterated analogs of water** and **chloroform**, unless you're absolutely sure your compound will dissolve in them. Always use the protonic solvent—H_2O or $CHCl_3$ here—for the solubility test. There are other deuterated solvents, and they may or may not be available for use. Check with your instructor.

SOME NMR INTERPRETATION

I've included a spectrum of ethylbenzene (Fig. 170) to give you some idea of how to start interpreting NMRs. Obviously, you'll need more than this. See your instructor or any good organic chemistry text for more information.

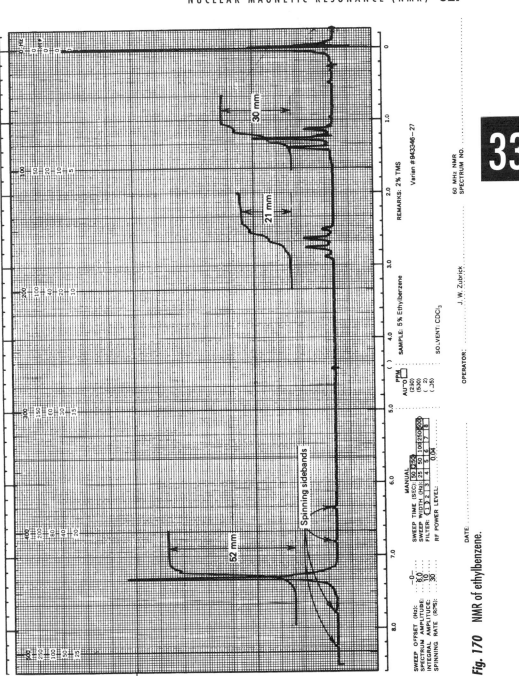

Fig. 170 NMR of ethylbenzene.

The Zero Point

Look at the extreme right of the NMR. That single, sharp peak comes from the protons in the **internal standard**, TMS. This signal is *defined as zero*, and all other values for the **chemical shift** are taken from this point. The units are parts per million (ppm) and use the Greek letter delta δ: $\delta 0.0$.

Protons of almost all other compounds you'll see will give signals *to the left of zero*; positive δ values, **shifted downfield** from TMS. There are compounds that give PMR signals **shifted upfield** from TMS: negative δ values.

Upfield and downfield are directions relative to where you point your finger on the NMR chart. Signals to the *right* of where you are: **upfield**. Signals to the *left* of where you are: **downfield**.

The Chemical Shift

You can see that not all the peaks fall in the same place, so the protons must be in different surroundings. There is one signal at $\delta 1.23$, one at $\delta 2.75$, and another at $\delta 7.34$. You usually take these values from the *center* of a **split signal** (that's coming up). See that the TMS is really zero before you report the chemical shift. If it is not at zero, you'll have to add or subtract one correction to all the values. This is the same as using a polystyrene calibration peak to get an accurate fix on IR peaks.

You'll need a correlation table or a correlation chart (Fig. 171) to help interpret your spectrum. The—CH_3 group is about in the right place ($\delta 1.23$). The $\delta 7.34$ signal is from the aromatic ring, and, sure enough, that's where signals from aromatic rings fall. The $\delta 2.75$ signal from the —CH_2— is a bit trickier to interpret. The chart shows a —CH_3 on a benzene ring in this area. Don't be literal and argue that *you don't have a* —CH_3; you have a —CH_2— CH_3. All right, they're different. But the —CH_2— group *is on a benzene ring and attached to a* —CH_3. That's why those —CH_2 protons are further downfield; that's why you don't classify them with ordinary R—CH_2—R protons. Use some sense and judgment.

I've blocked out related groups on the correlation table. Look at the set from $\delta 3.1$ to $\delta 4.0$. They're the area that protons on carbons attached to halogens fall in. Read that again. It's protons on *carbons* attached to halogens. The more electronegative the halogen on the carbon, the further *downfield* the *chemical shift of those protons*. The electronegative halogen

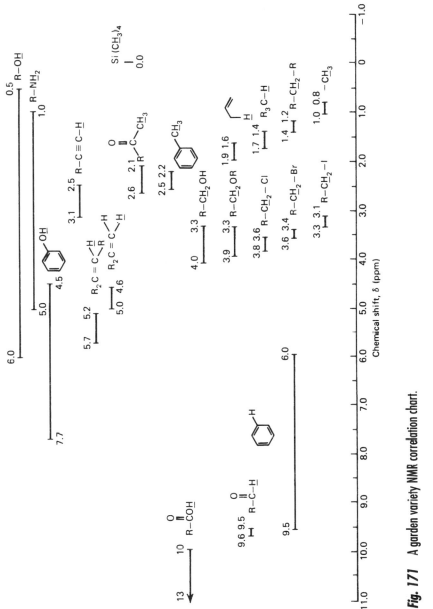

Fig. 171 A garden variety NMR correlation chart.

draws electrons from the carbon and thus from around the protons on the carbon. These protons now don't have as many electrons surrounding them. They are *not as shielded* from the big bad magnetic field as they might be. They are **deshielded**, so their signal falls *downfield*.

The hydrogen-bonded protons wander all over the lot. Where you find them, and how sharp their signals are, depend at least on the solvent, the concentration, and the temperature.

Some Ansisotropy

So what about aromatic protons ($\delta 6.0$–9.5), aldehyde protons ($\delta 5$–9.6), or even protons on double, nay triple, bonds ($\delta 5$–3.1)? All these protons are attached to carbons with π bonds, double or triple bonds, or aromatic systems. The electrons in these π bonds *generate their own little local magnetic field*. This local field in not *spherically symmetric*; it can shield or deshield protons depending on where the protons are—it's **anisotropic**. In Fig. 172, the **shielding regions** have pluses on them, and **deshielding regions** have minuses.

This is one of the quirks in the numbering system. Physically and psychosocially, a minus means *less (less shielding)*, and DOWNfield is further left on the paper; yet the value of δ goes UP. Another system uses

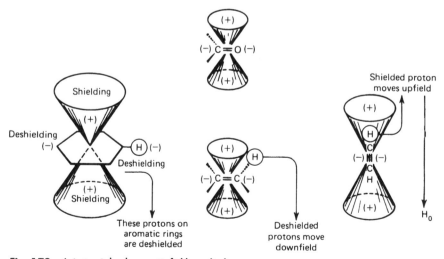

Fig. 172 Anisotropic local magnetic fields on display.

the Greek tau (τ)—that's 10.0–δ. So $\delta 0.00$ (ppm) is 10.0τ. Don't confuse these two systems. And don't ever confuse deshielding (or shielding) with the *proper direction of the chemical shift*.

Spin-Spin Splitting

Back to ethylbenzene. You'll find that the —CH_2— and the CH_3 protons are not single lines; they are *split*. **Spin-spin splitting**. Such a fancy name. Protons have a spin of plus or minus ½. If I'm sitting on the methyl group, I can see two protons on the adjacent carbon (—CH_2—). (**Adjacent carbon**, remember that.) They spin, so they produce a magnetic field. Which way do they spin? That's the crucial point. Both can spin one way, *plus*. Both can spin one way, *minus*. Or, each can go a different way; one plus, one minus.

Over at the methyl group (**adjacent carbon**, eh?), you can feel these fields. They add a little, they subtract a little, they cancel a little. So your methyl group splits into three peaks! It's split by the *two* protons *on the adjacent carbon*.

Don't confuse this with the fact that there are three protons on the methyl group! That Has Nothing To DO With It! It is mere coincidence.

The methyl group shows up as a **triplet** because it is split *by two* protons on the *adjacent carbon*.

Now what about the intensities? Why is the middle peak larger? Get out a marker and draw an A on one proton and a B on the other. OK. There's only one way for A and B to spin in the same direction: *Both A and B are plus, or both A and B are minus*. But there are *two ways* for them to spin opposite each other: *A plus with B minus; B plus with A minus*. This condition happens *two times*. Both A and B plus happen only *one time*. Both A and B minus happen only *one time*. So what? So the **ratio of the intensities is 1:2:1**. Ha! You got it—a **triplet**. Do this whole business sitting on the —CH_2— group. You get a **quartet**—*four lines*—*because the* —CH_2—*protons are adjacent to a methyl group*. They are split BY *three to give FOUR lines* (Fig. 173).

No, that is not all. You can tell that the —CH_2—protons and the —CH_3 protons split each other by their **coupling constant**, the distance between the split peaks of a single group. Coupling constants are called **J values** and are usually given in hertz (Hz). You can read them right from the chart, which has a grid calibrated in hertz. If you find protons at different

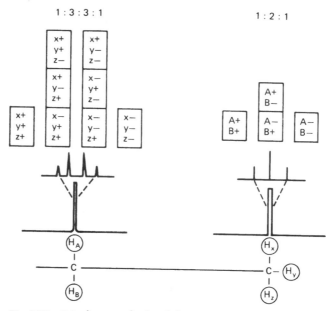

Fig. 173 Spin alignments for the ethyl group.

chemical shifts and their coupling constants are the same, they're splitting and coupling with each other.

Integration

Have you wondered about those funny curves drawn over the NMR peaks? They're **electronic integrations,** and they can tell you how many protons there are at each chemical shift. Measure the distances between the horizontal lines just before and just after each group. With a cheap plastic ruler, I get 52 mm for the benzene ring protons, 21 mm for the —CH_2— protons, and about 30 mm for the—CH_3 protons. Now you divide all the values by the smallest one. Well, 21 mm is the smallest, and without a calculator I get 2.47:1:1.43. Not even close. And how do you get that 0.47 or 0.43 proton? Try for the simplest *whole number ratio*. Multiply everything by 2, and you'll have 4.94:2:2.86. This is very close to 5:2:3, the actual number of protons in ethylbenzene. Use other whole numbers; the results are not as good, and you can't justify the splitting pattern—3 *split* BY 2 and 2 *split* BY 3—with other ratios. Don't use each piece of information in a vacuum.

There are a lot of other things in a typical NMR. There are **spinning sidebands**, small duplicates of stronger peaks, evenly spaced from the parent peak. They fall at *multiples of the spin rate*, which here is about 30 Hz. Spin the sample tube faster and these sidebands move farther away; slow the tube and they must get closer.

Signals that split each other tend to lean toward each other. It's really noticeable in the triplet and even distorts the intensity ratio in the quartet somewhat. Ask your instructor or see another textbook if you have questions.

THEORY OF
DISTILLATION

34

Distillation is one of the more important operations you will perform in the organic chemistry laboratory. It is important that you understand some of the physical principles working during a distillation. Different systems require different treatments; the explanations that follow parallel the classifications of distillations given earlier in the book.

CLASS 1: SIMPLE DISTILLATION

In a simple distillation, you recall, you separate liquids boiling BELOW 150°C at 1 atm from

1. Nonvolatile impurities.
2. Another liquid boiling at least 25°C higher than the first. The liquids should dissolve in each other. If they do not, then you should treat the system like a steam distillation, and if you're going to steam distill, be sure to look at the discussion for Class 4: Steam Distillations. The boiling points should have a 25°C difference because you may assume the higher boiling component doesn't do anything but sit there during the distillation. Otherwise, you may have a two-component system, and you also need to look at the discussion for Class 3: Fractional Distillation, as well as the discussion here.

That said, let's go on with our discussion of the distillation of one-component systems.

Suppose you prepared isobutyl alcohol by some means, and at the end of the reaction you wound up with a nice yellow-brown liquid. Checking any handbook, you find that isobutyl alcohol is a colorless liquid that boils at 108.1°C at 760 torr. On an STP day, the atmospheric pressure (P) is 760 torr and the boiling point is the **normal boiling point**. At that point, the single point when the vapor pressure of the liquid is the same as that of the atmosphere, the liquid boils.

Fearing little, you set up a Class 1, Simple Distillation and begin to heat the mix. If you kept track of the temperature of the *liquid* (and you don't; the thermometer bulb is up *above* the flask) and its *vapor pressure*, you'd get the temperature-vapor pressure data in columns 1 and 2 of Table 1.

At 82.3°C the vapor pressure of the liquid is 290.43 torr—much *lower* than atmospheric pressure. The liquid doesn't boil. Heat it to, say, 103.4°C, and the vapor pressure is 644.16 torr. Close to atmospheric pressure but no prize. Finally, at 108.1°C we have 760.04 torr. *The vapor pressure of the liquid equals that of the atmosphere, and the liquid boils.*

Now, do you see an entry in the table for brown gunk? Of course not. The brown gunk *must* have a very low (or no) vapor pressure at any temperature you might hit during your distillation. Without a vapor pressure, there can be no vapor. No vapor and there's nothing to condense. Nothing to condense, and there's no distillation. So the isobutyl alcohol comes over clean and pure, and the brown gunk stays behind.

If you plot the temperature and vapor pressure data given in Table 1, you reconstruct the **liquid-vapor equilibrium line** in the **phase diagram** of that liquid (Fig. 174). The equation of this line, and you might remember this from your freshman chemistry course, is the *Clausius–Clapeyron* equation:

$$p = p^0 \exp\left\{ \frac{-\Delta H}{R} \left(1/T - 1/T^0 \right) \right\}$$

Clausius and Clapeyron

So if you want to know how the vapor pressure of a substance is going to vary with temperature, you can use the *Clausius–Clapeyron* equation:

Table 1 Temperature–Vapor Pressure Data for Isobutyl and Isopropyl Alcohols.

Temperature (C°)	Isobutyl alcohol (torr)	Isopropyl alcohol (torr)
82.3	290.43	**760.00**
83.2	301.05	786.31
85.4	328.41	853.90
86.9	348.27	902.76
88.7	373.46	964.52
90.0	406.34	1044.83
95.8	488.59	1244.30
99.9	567.96	1435.11
103.4	644.16	1616.99
106.2	711.23	1776.15
108.1	**760.04**	1891.49

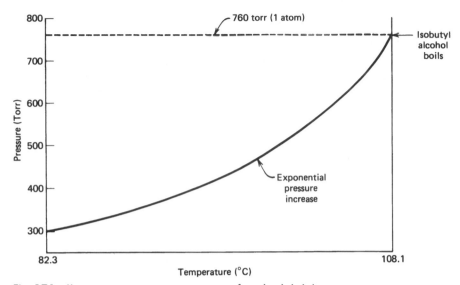

Fig. 174 Vapor pressure versus temperature curve for isobutyl alcohol.

$$p = p°\exp\left\{\frac{-\Delta H}{R}(1/T - 1/T°)\right\}$$

where

$p°$ is a known vapor pressure
$T°$ is the known temperature (in K, *not* °C)
These are usually taken from the normal boiling point.
 H is the heat of vaporization of the liquid.
R is the universal gas constant (8.314 J/mole–K).
T is the temperature you want the vapor pressure for.
p is the vapor pressure that you calculate for the temperature you
    want, T.

I've put this formula to two uses:

1. From the normal boiling point of isobutyl alcohol ($T° = 108.1$°C, 381.2K;
   $p°$=760 torr) and one other vapor pressure measurement that I found
   while doing research for this section (T=100°C, 373K; p=570 torr), I've
   gotten the **heat of vaporization (H).** This is the heat needed to
   vaporize a mole of pure isobutyl alcohol itself. Using these two values

in the Clausius–Clapeyron equation, I found that the H for isobutyl alcohol is 10039.70 cal/mole.

2. Now, with the H for isobutyl alcohol, I've calculated the field of pressures you see in Table 1 from temperatures of 82.3 to 108.1°C. *That's why the last pressure at 108.1 is 760.04, and not 760.00. The 760.04 value has been back-calculated using the H that we calculated from the top pressure-temperature points in the first place.*

I've also listed the vapor pressure data for isopropyl alcohol in column 3 of Table 1. Again, two steps were required to generate the data:

1. Two known vapor pressure-temperature points (T°=82.3°C, 355.4K, p°=760 torr; T=100°C, 373K, p=1440 torr) were used to calculate the H:9515.73 cal/mol.
2. Now that H for isopropyl alcohol, and temperatures from 82.3 to 108.1°C, were used to calculate a field of vapor pressures for isopropyl alcohol. These are in column 3 of Table 1.

I've done these calculations for several good reasons:

1. To show you how to use the Clausius–Clapeyron equation, and to show you how well the equation fits over small temperature ranges. The calculated boiling point pressure for isobutyl alcohol (760.04 torr) is not very different from the normal boiling point pressure of 760.00 torr (0.005%).
2. To show you that compounds with *higher H's have lower vapor pressures*. This means that it takes more energy to vaporize them.
3. To show you that you can calculate vapor pressures that are *above* the boiling point of the liquid. They have a slightly different meaning, however. There is no liquid isopropyl alcohol at 100°C and 760 torr. The vapor pressure at 100°C is 1440 torr, almost twice the atmospheric pressure. But if we artificially increased the pressure over a sample of isopropyl alcohol (pumped up the flask with compressed air?) to 1440 torr and then heated the flask, the alcohol would no longer boil at 82.3°C. You'd have to go as high as—did someone say 100°C—before the vapor pressure of the liquid matched the now pumped-up atmospheric pressure,and the liquid would boil.
4. To show you the theory of the next topic, Class 3: Fractional Distillation.

CLASS 3: FRACTIONAL DISTILLATION

In a fractional distillation, you remember, you are usually separating liquid mixtures, soluble in one another, that boil at less that 25°C from each other at a pressure of one atmosphere.

Now I didn't discuss *both* isopropyl and isobutyl alcohol in the last section for my health. Suppose you're given a mixture of these two to separate. They *are* miscible in each other, and their boiling points are just about 25°C apart. A textbook case, eh?

A Hint from Dalton

So you set up for a fractional distillation and begin to heat the liquid mixture. After a bit, it boils. And what does that mean? The vapor pressure of the *solution* (not just one component) is now equal to the atmospheric pressure, 760 torr. (We're very lucky that textbook-land has so many STP days.) Each component exerts its own vapor pressure, and when the *total* pressure reaches 760 torr, the solution boils.

$$P_{Total} = P_A + P_B$$

This is **Dalton's law of partial pressures.** The total pressure of the gang is equal to the sum of their individual efforts. Here, A could be the isopropyl alcohol and B the isobutyl (it doesn't matter), but P_{Total} must be the *atmospheric pressure*, P_{atm}. So a special version of Dalton's law of partial pressures for use in fractional distillation will be

$$P_{atm} = P_A + P_B$$

Dalton and Raoult

If that's all there were to it, we'd be talking about Class 4: Steam Distillations and the like, where the components aren't soluble, and we could quit. Here, *the two are soluble in each other*. The individual vapor pressures of each component (P_A, P_B) depend not only on the temperature, but also on their **mole fraction**.

It makes sense, really. Molecules are the beasties escaping from solution during boiling, and, well, if the two liquids dissolve in each other perfectly, the more molecules (moles) of one component you have, the more the solution behaves like that one component, until it gets to be the same as a one-component liquid. This is Raoult's law

$$P_A = X_A P_A^\circ$$

where

P_A is the vapor pressure of A from the mixture
X_A is the mole fraction of liquid A
P°_A is the vapor pressure of the pure liquid A

If we change the As to Bs, can you still follow me? It's the same thing, only now with liquid B. If we combine the special case of Dalton's law with Raoult's law, we get

$$P_{atm} = X_A P_A^\circ + X_B P_B^\circ$$

Look at this. If there is NO B, then the fraction of A is 1, and the pure liquid A boils when its vapor pressure equals the atmospheric pressure. Didn't I say that? Similarly, for B without A, the mole fraction of B is 1, and it too boils when its vapor pressure equals the atmospheric pressure.

At this point, you're usually given the temperature versus mole fraction diagram for two miscible liquids (Fig. 175), and you're told it's a consequence of Raoult's law. Well, yes. But not directly. Raoult's law is a relationship of *pressure*, not temperature, versus mole fraction, and this law is pretty much straight line. You don't need all your orbitals filled to see that you've been presented with a *temperature versus mole fraction* diagram, there are *two lines* (not one), and neither of them is very straight.

A Little Algebra

I want to convert the combined laws of Dalton and Raoult so that I can show the variation in mole fraction explicitly. First, you'll agree that the mole fractions of A and B *must* add to 1 (or they wouldn't be fractions, eh?), so

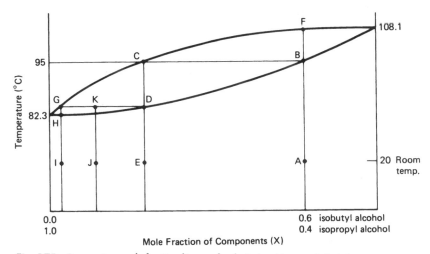

Fig. 175 Temperature-mole fraction diagram for the isobutyl-isopropyl alcohol systems.

$$X_A = X_B = 1$$

Now back with the Dalton and Raoult

$$P_{atm} = X_A P_A^\circ + X_B P_B^\circ$$

and seeing that $X_B = 1 - X_A$, I substitute to get

Expand this expression with a multiplication:

$$P_{atm} = X_A P_A^\circ + P_B^\circ - X_A P_B^\circ$$

Collect the terms with the mole fraction in them:

$$P_{atm} = X_A P_A^\circ - X_A P_B^\circ + P_B^\circ$$

And factor the mole fraction out to give:

$$P_{atm} = X_A \left(P_A^{\circ} - P_B^{\circ} \right) + P_B^{\circ}$$

To isolate the mole fraction X_A subtract P_B° from both sides:

$$P_{atm} - P_B^{\circ} = X_A \left(P_A^{\circ} - P_B^{\circ} \right)$$

and divide by $\left(P_A^{\circ} - P_B^{\circ} \right)$ to get

$$X_A = \frac{P_{atm} - P_B^{\circ}}{\left(P_A^{\circ} - P_B^{\circ} \right)}$$

Clausius and Clapeyron Meet Dalton and Raoult

We still have a formula that relates mole fraction to pressure. Note, however, that with the exception of the atmospheric pressure (P_{atm}) *all* the other pressures are those of pure liquids (and P_B°). Now, how does the vapor pressure of a pure liquid vary with temperature? Smite your forehead and say that you could have had a V-8. The *Clausius–Clapeyron* equation:

$$p = p^{\circ} \exp \left\{ \frac{-\Delta H}{R} (1/T - 1/T^{\circ}) \right\}$$

The P°'s of Dalton–Raoult are vapor pressures taken at fixed temperatures. They are the p's in the Clausius–Clapeyron equation found with a variation in temperature. You don't believe me? Pick a vapor pressure and temperature pair from Table 1 for either liquid, and let these be p° and T° (and don't forget to use K, not °C). Now what happens when the "unknown" temperature (T) is the same as T°? The ($1/T$–$1/T^{\circ}$) becomes zero, the entire exponent becomes zero, p° is multiplied by 1 (anything to the power zero is 1, eh?), and so on $p=p^{\circ}$.

The next part is messy, but somebody's got to do it. I'm going to use the vapor pressure-temperature data for the normal boiling points of both liquids in the Clausius–Clapeyron equation. Why? They're convenient, *known* vapor pressure-temperature points. When I do this, however, I exercise my right to use different superscripts to *impress on you that these points are the normal boiling points*. So for liquid A, we have and ; if A is isobutyl alcohol, =760 torr and =101.8°C. For liquid B, we have and ; if B is isopropyl alcohol, = 760 torr and =82.3°C.

Using the new letters on the previous page and substituting the Clausius–Clapeyron equation for every P° you get

$$\frac{P_{atm} - P_B^* \exp\left\{\frac{-\Delta H_B}{R}\left(1/T - 1/T_B^*\right)\right\}}{P_A^* \exp\left\{\frac{-\Delta H_A}{R}\left(1/T - 1/T_A^*\right)\right\} - P_B^* \exp\left\{\frac{-\Delta H_B}{R}\left(1/T - 1/T_B^*\right)\right\}}$$

Phew!

The only variables in this beast are the mole fraction (X) and the temperature (T). Every other symbol is a constant. We finally have the equation for the bottom line of the temperature-mole fraction diagram, something that has eluded us for years.

Dalton Again

What about the upper curve? Glad you asked (sigh). The composition in the vapor is also related to Dalton's law of partial pressures. For an ideal gas $PV = nRT$ (and you thought you'd never see that again!) and for the vapors above the liquid,

$$P_A = n_A \frac{RT}{V} \text{ and } P_B = n_B \frac{RT}{V}$$

Yet we know that the total pressure of Dalton's gang is the sum of their individual efforts:

$$P_{Total} = P_A + P_B$$

So,

$$P_{Total} = n_{Total}\left(\frac{RT}{V}\right)$$

Now watch, as I divide the pressure of A by the total pressure:

$$P_A / P_{Total} = n_A / n_{Total} \left(\frac{\frac{RT}{V}}{\frac{RT}{V}} \right)$$

Well, the RT/Vs cancel giving

$$P_A / P_{Total} = n_A / n_{Total}$$

The ratio of the number of moles of A to the *total* number of moles is the **mole fraction of component A in the vapor**

$$X_A^{Vapor} = P_A / P_{Total}$$

P_{Total} for the ordinary distillation is the atmospheric pressure, 760 torr, (P_B's, eh?). P_A is the vapor pressure of A, and again by Raoult's Law, $P_A = X_A P_A^\circ$. Putting the two together, we get

$$X_A^{Vapor} = X_A^{Liquid} P_A^\circ / 760$$

We can make the same kind of substitution of Clausius–Clapeyron here, and we get a similarly curved function for the upper line in the temperature-mole fraction diagram.

To show you that all this really does work, I've listed the experimental composition data for the isopropyl/isobutyl alcohol system from Landolt-Bornstein (Landolt-Bornstein is to physical chemistry what Beilstein is to organic. And wouldn't that make for a wild analogy question on the college board entrance exams?), along with my calculated data (Table 2). (That explains my choice of temperatures for Table 1.) I've also given the absolute and percent differences between the experimental data, and what I've calculated. These differences are on the order of 1% or less, a very good agreement indeed.

So now, for any two liquids, if you have their normal boiling points and vapor pressures at any other temperature, you can generate the temperature-mole fraction diagram.

Table 2 Experimental and Calculated Data.

Temperature vs. Mole Fraction Calculations

Compound	Normal BP (C)	Vapor Pressure @ 100 (C)
2-Propanol	82.3	1440 torr
2-Methyl-1-propanol	108.2	570 torr

Data from Moore (3rd. ed.)

Calculated Δ H (VAP)	2-Propanol	9515.73 cal/mol
	2-Methyl-1-propanol	10039.70 cal/mol

Comparison of Data from Moore and Calculated Data (T=100 (C))

		Moore	Calc.
Vapor pressures:	2-Propanol	1440	1440.05
(torr)	2-Methyl-1-propanol	570	570.02
Mole fraction:	2-Propanol	0.219	0.2184
(in liquid)	2-Methyl-2-propanol	0.781	0.7861
Mole fraction:	2-Propanol	0.415	0.4137
(in vapor)	2-Methyl-2-propanol	0.585	0.5863

Mole Fraction of 2-Propanol Liquid Data

#	T(C)	X Calc	X Lit.	Diff	% Diff
1	83.2	0.9458	0.9485	−.0027	−0.286
2	85.4	0.8213	0.8275	−.0062	−0.748
3	86.9	0.7425	0.7450	−.0025	−0.331
4	88.7	0.6540	0.6380	0.0160	2.505
5	90.9	0.5539	0.5455	0.0084	1.540
6	95.8	0.3591	0.3455	0.0136	3.949
7	99.9	0.2215	0.2185	0.0030	1.357
8	103.4	0.1191	0.1155	0.0036	3.096
9	106.2	0.0458	0.0465	−.0007	−1.507

Mole Fraction of 2-Propanol Vapor Data

#	T(C)	X Calc	X Lit.	Diff	% Diff
1	83.2	0.9785	0.9805	−.0020	−0.201
2	85.4	0.9228	0.9295	−.0067	−0.723
3	86.9	0.8820	0.8875	−.0055	−0.618
4	88.7	0.8300	0.8245	0.0055	0.663
5	90.9	0.7615	0.7580	0.0035	0.460
6	95.8	0.5880	0.5845	0.0035	0.600
7	99.9	0.4182	0.4270	−.0088	−2.063
8	103.4	0.2533	0.2510	0.0023	0.936
9	106.2	0.1070	0.1120	−.0050	−4.433

What Does It All Mean?

Getting back to the temperature-mole fraction diagram (Fig. 175), suppose you start with a mixture such that the mole fractions are as follows: isobutyl alcohol, 0.60, and isopropyl alcohol, 0.40. On the diagram, that composition is point A at a room temperature of 20°C. Now you heat the mixture, and you travel upward from point A to point B. The liquid has the same composition; it's just hotter.

At 95°C, point B, the mixture boils. Vapor, with the composition at point C, comes flying out of the liquid. (The horizontal line tying the composition of the vapor to the composition of the liquid is the **liquid-vapor tie-line.**), and this vapor condenses (point C to point D), say, part of the way up your distilling column.

Look at the composition of this *new liquid* (point E). It is *richer in the lower-boiling component*. The step cycle B-C-D represents one distillation.

If you heat this new liquid that's richer in isopropyl alcohol (point D), you get vapor (composition at point G along a horizontal tie-line) that condenses to liquid H. So step cycle D-G-H is another distillation. The two steps represent two distillations.

Much of this work was carried out using a special distilling column called a **bubble-plate column** (Fig. 176). Each *plate* really does act like a distilling flask with a very efficient column, and one distillation is really carried out on one physical plate. To calculate the number of plates (separation steps or distillations) for a bubble-plate column, you just count them!

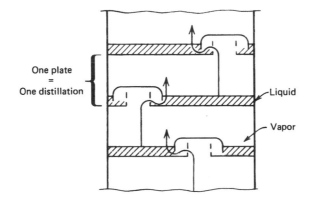

Fig. 176   A bubble-plate fractionating column.

Unfortunately, the fractionating column you usually get is not a bubble-plate type. You have an open tube that you fill with column packing (see "Class 3: Fractional Distillation" in Chapter 20) and *no plates*. The distillations up this type of column are not discrete, and the question of where one plate begins and another ends is meaningless. Yet, if you use this type of column, you do get a better separation than if you used no column at all. It's as *if* you had a column with some bubble-plates. And if your distilling column separates a mixture as well as a bubble-plate column with two real plates, you must have a column with two **theoretical plates.**

You can calculate the number of theoretical plates in your column if you distill a two-component liquid mixture of known composition (isobutyl and isopropyl alcohol perhaps?) and collect a few drops of the liquid condensed from the vapor at the top of the column. You need to determine the composition of that condensed vapor usually from a calibration curve of known compositions versus their refractive indices [see Chapter 28, "Refractometry"], and you must have the temperature-mole fraction diagram (Fig. 171).

Suppose you fractionated that liquid of composition A, collected a few drops of the condensed vapor at the top of the column, analyzed it by taking its refractive index, and found that this liquid had a composition corresponding to point J on our diagram. You would follow the same path as before (B-C-D, one distillation; D-G-H, another distillation) and find that composition falls a bit short of the full cycle for distillation #2.

Well, all you can do is estimate that it falls at, say, a little more than halfway along this second tie-line, eh (point K)? OK then. This column has been officially declared to have 1.6 theoretical plates. Can you have tenths of plates? Not with a bubble-plate column, but certainly with any column that does not have discrete separation stages.

Now you have a column with one-point-six theoretical plates. "Is that good?" you ask. "Relative to what," I say. If that column is 6 feet high, that's terrible. The Height Equivalent to a Theoretical Plate (HETP) is 3.7 feet/plate. Suppose another column also had 1.6 theoretical plates, but was only 6 inches (0.5 feet) high. The HETP for this column is 3.7 inches/plate, and *if it were 6 feet high*, it would have 19 plates. The *smaller* the HETP, the more efficient the column is. There are more plates for the same length.

One last thing. On the temperature-mole fraction diagram, there's a point F that I haven't bothered about. F is the grade you'll get when you extend the A-B line up to cut into the upper curve and you then try to do anything with this point. I've found an amazing tendency for some people to extend that line to point F. Why? Up the temperature from A to B, and the sample boils. *When the sample boils, the temperature stops going up.* Heat going into the distillation is being used to vaporize the liquid (heat of vaporization, eh?) and all you get is a vapor, enriched in the lower-boiling component, with the composition found at the end of a *horizontal* tie-line.

Reality Intrudes I: Changing Composition

To get the number of theoretical plates, we fractionally distilled a known mixture and took off a small amount for analysis, so as not to disturb things very much. You, however, have to fractionally distill a mixture and hand in a good amount of product, and do it within the time limits of the laboratory.

So when you fractionally distill a liquid, you *continuously* remove the lower-boiling fraction from the *top* of the column. And where did that liquid come from? The boiling fluid at the *bottom* of the column. Now if the distillate is richer in the lower-boiling component, what happened to the composition of the boiling liquid? I'd better hear you say that the boiling liquid gets richer in the *higher-boiling* component (Fig. 177).

So as you fractionally distill, not only does your boiling liquid get richer in the higher component, but so also does your distillate, your condensed vapor. Don't worry too much abut this effect. It happens as long as you have to collect a product of revaluation. Let your thermometer be your guide, and keep the temperature spread less than 2°C.

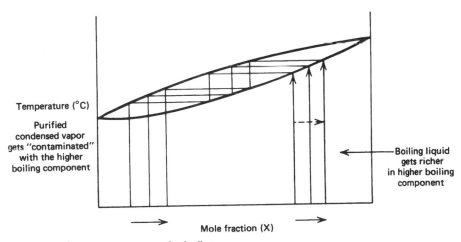

Fig. 177 Changing composition as the distillation goes on.

Reality Intrudes II: Nonequilibrium Conditions

Not only were we forced to remove a small amount of liquid to accurately determine the efficiency of our column, but also we had to do it *very slowly*. This allowed the distillation to remain *at equilibrium*. The **throughput**, the rate at which we took material out of the column, was *very low*. Of all the molecules of vapor that condensed at the top of the column, *most fell back down the column; few were removed*. A very high **reflux ratio**. With an infinite reflux ratio (no throughput), the condensed vapor at the top of the column is as rich in the lower-boiling component as it's ever likely to get in your setup. As you remove this condensed vapor, the equilibrium is upset, as more molecules rush in to take the place of the missing. The faster you distill, the less time there is for equilibrium to be reestablished and the less time there is for the more volatile components to sort themselves out and move to the top of the column. So you begin to remove higher-boiling fractions as well, and you cannot get as clean a separation. In the limit, you could remove vapor so quickly that you shouldn't have even bothered using a column.

Reality Intrudes III: Azeotropes

Occasionally, you'll run across liquid mixtures that cannot be separated by fractional distillation. That's because the composition of the vapor coming off the liquid is the *same* as that of the liquid itself. You have an **azeotrope**, a liquid mixture with a constant boiling point.

Go back to the temperature-mole fraction diagram for the isopropyl alcohol-isobutyl alcohol system (Fig. 175). The composition of the vapor is *always different from* that of the liquid, and we can separate the two compounds. If the composition of the vapor is the same as that of the liquid, that separation is hopeless. Since we've used the notions of an ideal gas in deriving our equations for the liquid and vapor compositions (Clausius–Clapeyron, Dalton, and Raoult), this azeotropic behavior is said to result from *deviation from ideality*, specifically *deviations from Raoult's law*. Although you might invoke certain interactive forces in explaining nonideal behavior, you cannot predict azeotrope formation *a priori*. Very similar materials form azeotropes (ethanol–water). Very different materials form azeotropes (toluene–water). And they can be either **minimum-boiling azeotropes** or **maximum-boiling azeotropes**.

Minimum-Boiling Azeotropes

The ethanol-water azeotrope (95% ethanol–5% water) is an example of a minimum-boiling azeotrope. Its boiling point is lower than that of the components (Fig. 178). If you've ever fermented anything and distilled the results in the hopes of obtaining 200 proof (100%) white lighting, you'd have to content yourself with getting the azeotropic 190 proof mixture instead. Fermentation usually stops when the yeast die in their own 15% ethanol solution. At room temperature, this is point A on our phase diagram. When you heat the solution, you move from point A to point B and, urges to go to point F notwithstanding, you cycle through distillation cycles B-C-D and D-E-G. And, well, guess what comes off the liquid? Yep, the azeotrope. As the azeotrope comes over, the composition of the boiling liquid moves to the right (it get richer in water), and finally there isn't enough ethanol to support the azeotropic composition. At that point, you're just distilling water. The process is mirrored if you start with a liquid that is >95% ethanol and water. The azeotrope comes off first.

Maximum-Boiling Azeotropes

The chloroform-acetone azeotrope (52% chloroform–48% acetone) is an example of the much rarer maximum-boiling azeotrope. Its boiling point is higher than that of the components (Fig. 179). At compositions off the

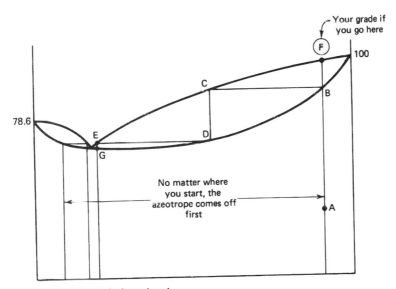

Fig. 178   Minimum-boiling ethanol–water azeotrope.

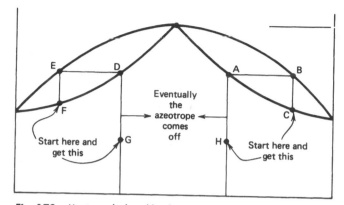

Fig. 179 Maximum-boiling chloroform-acetone azeotrope.

azeotrope, you do distillations A-B-C or D-E-F (and so on) until the boiling liquid composition reaches the azeotropic composition. Then that's all that comes over. So, initially, one of the components comes off first.

Azeotropes on Purpose

You might think the formation of azeotropes is an unalloyed nuisance, but they can be useful. Toluene and water form a minimum-boiling azeotrope (20.2% water; 85°C). If for some reason, you need *very dry* toluene, all you have to do is distill some of it. The water–toluene azeotrope comes off first, and, well, there goes all the water. It gets removed by **azeotropic distillation**. The technique can also be used in certain reactions, including the preparation of amides from very high-boiling acids and amines. (They can even be solids.) You dissolve the reagents in toluene, set up a reflux condenser fitted with a **Dean-Stark trap** (Fig. 180), and let the mixture reflux. As the amide forms, water is released and the water is constantly removed by azeotropic distillation with the toluene. The azeotrope cools, condenses, and collects in the Dean-Stark trap. At room temperature, the azeotrope is said to "break," and the water forms a layer at the bottom of the trap. Measure the amount of water, and you have an idea of the extent of the reaction.

Absolute (100%) ethanol is often made by adding benzene to the ethanol-water **binary azeotrope** (two components) to make a **ternary azeotrope** (three components). This ternary alcohol-water-benzene (18.5:7.4:74.1) azeotrope comes over until all the water is gone, followed by a benzene-ethanol mixture. Finally, absolute ethanol gets its chance to appear, marred only slightly by traces of benzene.

Other Deviations

The furfural-cyclohexane phase diagram (Fig. 181) shows that you can have mixtures that exhibit nonideal behavior without having to form an azeotrope. In sum, without the phase diagram in front of you, you shouldn't take the distillation behavior of any liquid mixture for granted.

34

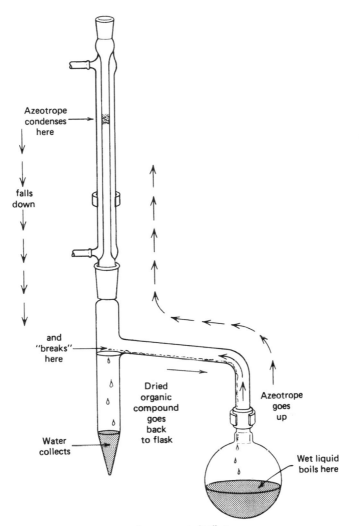

Fig. 180 Removing water by azeotropic distillation.

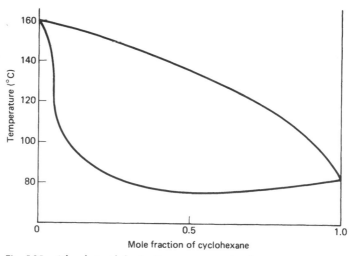

Fig. 181 Other deviant behavior (but no azeotropes) in the furfural–cyclohexane system.

CLASS 4: STEAM DISTILLATION

Steam distillation is for the separation of mixtures of tars and oils, and they must not dissolve much in water. If you think about it a bit, this could be considered a fractional distillation of a binary mixture with an *extreme* deviation from Raoult's law. The water and the organic oils want nothing to do with each other. So much so, that you can consider them unmixed, in separate compartments of the distilling flask. As such, they act completely independently of each other. The mole fraction of each component in its own compartments is 1. So Raoult's Law becomes

$$P_{\text{Total}} = P_A + P_B$$

This is just Dalton's Law of partial pressures. P_{Total} is P_{atm} for a steam distillation. So the vapor pressure of the organic oil is now less than that of the atmosphere and the water, and codistills at a much lower temperature.

As an example, suppose you were to try to directly distill quinoline. Quinoline has a boiling point of 237°C at 1 atm. Heating organic molecules to these temperatures may often be a way to decompose them. Fortunately, quinoline is insoluble in water, and it does have some vapor pressure at about the boiling point of water (10 torr at 99.6°C). If it had a *much lower* vapor pressure at the boiling point of water (say, 0.1 torr), there couldn't be enough of it vaporizing to make even steam distillation worthwhile.

Well, at 99.6°C, quinoline contributes 10 torr to the total vapor pressure, and water must make up the difference (750 torr) in order to satisfy Dalton's Law of partial pressures to make P_{Total} 760 torr at boiling. Using the relationships for the composition of the vapor *over a liquid*, we can calculate the quinoline/water ratio coming over.

If we consider each to be an ideal gas, then

$$PV = nRT \text{ (yes, again)}$$

The number of moles (n) of anything is just the weight in grams (g) divided by the *molecular weight* of that substance (MW), and so

$$PV = (g \,/\, \mathrm{MW})(\mathrm{RT})$$

Multiplication by MW gives

$$(\mathrm{MW})PV = g(\mathrm{RT})$$

and isolating the mass of the material by dividing by RT gives

$$(\mathrm{MW})PV \,/\, \mathrm{RT} = g$$

Now for *two* vapors, A and B, I'll construct a ratio where

$$\frac{(\mathrm{MW})_A \, P_A V \,/\, \mathrm{RT} = g_A}{(\mathrm{MW})_B \, P_B V \,/\, \mathrm{RT} = g_B}$$

With A and B in the same flask, R, T, and V must be the same for each, and we can cancel these terms giving

$$\frac{(\mathrm{MW})_A \, P_A = g_A}{(\mathrm{MW})_B \, P_B = g_B}$$

If we plug in values for the molecular weights and vapor pressures of quinoline and water, we get

$$\frac{(129)(10)}{(18)(750)} = \frac{0.0956}{1}$$

So, as an approximation, for every $10\,g$ of distillate we collect, $1\,g$ will be our steam-distilled quinoline.

THEORY OF
EXTRACTION

"Several small extractions are better than one big one." Doubtless you've heard this many times, but now I'm going to try to show that it is true.

By way of example, let's say you have an *aqueous* solution of oxalic acid, and you need to isolate it from the water by doing an extraction. In your handbook, you find some solubilities of oxalic acid as follows: 9.5 g/100 g in water; 23.7 g/100 g in ethanol; 16.9 g/100 g in diethyl ether. Based on the solubilities, you decide to extract into ethanol from water, forgetting for the moment that ethanol is *soluble in water* and that you *must have two insoluble liquids* to carry out an extraction. Chagrined, you choose diethyl ether.

From the preceding solubility data we can calculate the **distribution,** or **partition coefficient** for oxalic acid in the water-ether extraction. This coefficient (number) is just the ratio of solubilities of the compound you wish to extract in the two layers. Here,

$$K_p = \frac{\text{solubility of oxalic acid in ether}}{\text{solubility of oxalic acid in water}}$$

which amounts to 16.9/9.5, or 1.779.

Imagine you have 40 g of oxalic acid in 1000 ml of water, and you put that in contact with 1000 ml of ether. The oxalic acid *distributes itself between the two layers.* How much is left in each layer? Well, if we let x g equal the amount that stays in the water, $1.779x$ g of the acid has to walk over to the ether. And so

$$\text{Wt of oxalic acid in ether} = (1000 \text{ ml})(1.779x \text{ g/ml}) = 1779x \text{ g}$$
$$\text{Wt of oxalic acid in water} = (1000 \text{ ml})(x \text{ g/ml}) = 1000x \text{ g}$$

The total weight of the acid is 40 g (now partitioned between two layers) and

$$2779x \text{ g} = 40 \text{ g}$$
$$x = 0.0144$$

and

$$\text{Wt of oxalic acid in ether} = 1779 (0.0144)\text{g} = 25.6 \text{ g}$$
$$\text{Wt of oxalic acid in water} = 1000 (0.0144)\text{g} = 14.4 \text{ g}$$

Now, let's start with the same 40 g of oxalic acid in 1000 ml of water, but this time we will *do three extractions with 300 ml of ether*. The first 300 ml portion hits, and

Wt of oxalic acid in ether = (300 ml)(1.779x g/ml) = 533.7x g
Wt of oxalic acid in water = (1000 ml)(x g/ml) = 1000x g

The total weight of the acid is 40 g (now partitioned between two layers) and

$$1533.7x \text{ g} = 40g$$
$$x = 0.0261$$

and

Wt of oxalic acid in ether = 533.7 (0.0261)g = 13.9 g
Wt of oxalic acid in water = 1000 (0.0261)g = 26.1 g

That ether layer is removed, and the *second jolt* of 300 ml fresh ether hits, and

Wt of oxalic acid in ether = (300 ml)(1.779x g/ml) = 533.7x g
Wt of oxalic acid in water = (1000 ml)(x g/ml) = 1000x g

But here we started with 26.1 g of acid in water (now partitioned between two layers) and

$$1533.7x \text{ g} = 26.1g$$
$$x = 0.0170$$

and

Wt of oxalic acid in ether = 533.7 (0.0170)g = 9.1 g
Wt of oxalic acid in water = 1000 (0.0170)g = 17.0 g

Again, *that ether layer is removed,* and the *third jolt* of 300 ml fresh ether hits, and

$$\text{Wt of oxalic acid in ether} = (300 \text{ ml})(1.779x \text{ g/ml}) = 533.7x \text{ g}$$
$$\text{Wt of oxalic acid in water} = (1000 \text{ ml})(x \text{ g/ml}) = 1000x \text{ g}$$

But here, we started with 17.0 g of acid in water (now partitioned between two layers) and

$$1533.7x \text{ g} = 17.0 \text{g}$$
$$x = 0.011$$

and

$$\text{Wt of oxalic acid in ether} = 533.7 \ (0.011)\text{g} = 5.87 \ g$$
$$\text{Wt of oxalic acid in water} = 1000 \ (0.011)\text{g} = 11.0 \ g$$

(They don't quite add up to 17.0 g—I've rounded them off a bit.)

Let's consolidate what we have. First, 13.9 g, then 8.5 g and, finally 5.34 g of oxalic acid, for a total of 27.7 g of acid extracted into 900 ml of ether. OK, that's not far from 24.7g extracted *once* into 1000 ml of ether. That's because the distribution coefficient is fairly low. But it is *more.* That's because *several small extractions are better than one large one.*

INDEX